An introduction to
THE MECHANICS OF SOILS AND FOUNDATIONS
Through Critical State Soil Mechanics

Books are to be returned on or

A

25 FEB 2000

AN INTRODUCTION TO THE MECHANICS OF SOILS AND FOUNDATIONS

An Introduction to the Mechanics of Soils and Foundations
is a new title in the
McGraw-Hill International Series in Civil Engineering

International Editorial Board

An Introduction to

THE MECHANICS OF SOILS AND FOUNDATIONS

Through Critical State Soil Mechanics

John Atkinson

Professor of Soil Mechanics
City University, London

McGRAW-HILL BOOK COMPANY

London · New York · St Louis · San Francisco · Auckland
Bogotá · Caracas · Lisbon · Madrid · Mexico · Milan
Montreal · New Delhi · Panama · Paris · San Juan
São Paulo · Singapore · Sydney · Tokyo · Toronto

Published by
McGRAW-HILL Book Company Europe
Shoppenhangers Road, Maidenhead, Berkshire, SL6 2QL, England
Telephone 0628 23432
Fax 0628 770224

British Library Cataloguing in Publication Data

Atkinson, J. H.
  Introduction to the Mechanics of Soils and Foundations. –
(McGraw-Hill International Series in Civil Engineering)
  I. Title II. Series   624.1
  ISBN 0–07–707713–X

Library of Congress Cataloging-in-Publication Data

Atkinson, J. H.
  An introduction to the mechanics of soils and foundations :
through critical state soil mechanics / John Atkinson.
    p.   cm.—(McGraw-Hill international series in civil
engineering)
  Includes bibliographical references and index.
  ISBN 0–07–707713–X
  1. Soil mechanics. 2. Foundations.   3. Slopes (Soil mechanics)
I. Title.   II. Series.
TA710.A786   1993
624.1'5136—dc20   92-37323 CIP

345 CL 9654

Typeset by Keyword Publishing Services, London
Printed and bound in Great Britain by Clays Ltd, St Ives plc

CONTENTS

This book is about the behaviour of engineering soils and simple geotechnical structures such as foundations and slopes and it covers most of the theoretical geotechnical engineering content of a degree course in civil engineering. The book is aimed primarily at students taking first degree courses in civil engineering but it should also appeal to engineers, engineering geologists and postgraduate students wishing for a simple and straightforward introduction to the current theories of soil mechanics and geotechnical engineering. Although it deals specifically with soils and soil mechanics many of the theories and methods described apply also to rocks and rock mechanics.

The teaching and practice of geotechnical engineering has undergone significant changes in the past 25 years or so, both in the development of new theories and practices and in the standing of the subject within the civil engineering curriculum. Geotechnical engineering is now regarded as one of the major disciplines in civil engineering analysis (the others being hydraulics and structures). The most important development, however, has been the unification of shearing and volumetric effects in soil mechanics in the theories known generally as critical state soil mechanics and application of these theories in geotechnical analysis. In this book, unlike most of the other contemporary books on soil mechanics, the subject is developed using the unified theories right from the start, and theories for stability of foundations and slopes are developed through the upper and lower bound plasticity methods as well as the more commonly used limit equilibrium method. This is an up-to-date approach to soil mechanics and geotechnical engineering and it provides a simple and logical framework for teaching the basic principles of the subject.

The term 'critical state soil mechanics' means different things to different people. Some take critical state soil mechanics to include the complete mathematical model known as Cam Clay and they would say that this is too advanced for an undergraduate course. My view is much simpler, and by critical state soil mechanics I mean the combination of shear stress, normal stress and volume into a single unifying framework. In this way a much clearer idea emerges of the behaviour of normally consolidated and overconsolidated soils during drained and undrained loading up to, and including, the ultimate or critical states. It is the relationship between the initial states and the critical states that largely determines soil behaviour. This simple framework is extremely useful for teaching and learning about soil mechanics and it leads to a number of simple analyses for stability of slopes, walls and foundations.

This book is based on courses of lectures given to undergraduate students in civil engineering at City University. In the first year students take a course in geology and they also take a course in mechanics of materials within which there are six to eight lectures on soil mechanics and geotechnical engineering. These lectures cover the whole of the conventional syllabus (classification, seepage, strength, consolidation, bearing capacity and settlement, slope stability and earth pressure) but at lightning speed. The object is to introduce the students to the concepts and vocabulary of geotechnical engineering within the context of conventional mechanics of materials and structures and with reference to their everyday, childhood experiences of playing with sand, flour, plasticine and other soil-like materials so that, as the course develops in later years, they can relate particular topics into the whole scheme of civil engineering.

In the second year the students take a major course of lectures (with several laboratory sessions) in theoretical soil mechanics and geotechnical engineering. This is based on my earlier books—*The Mechanics of Soils* (with Peter Bransby) and *Foundations and Slopes*. This course depends entirely on the unification of shearing and volumetric effects which is introduced right from the start (and had been in the first year), although the phrase 'critical state soil mechanics' is rarely used. Theoretical soil mechanics is taken up to the development of a complete state boundary surface but stops short of the mathematical treatment of Cam clay. Stability problems are solved using upper and lower bound methods and these are then used to introduce limit equilibrium methods and standard tables and charts for bearing capacity, slope stability and earth pressure. In the third year the course covers practical aspects of geotechnical engineering through a series of lectures and projects on topics such as ground investigation, foundations, slopes, retaining walls and embankment designs.

This book covers the material in the second-year course (and also that summarized in the first year). It does not deal specifically with the practical aspects of geotechnical engineering which are introduced in the third year and are, in any case, generally better learned through working in practice with experienced engineers. This book should provide the basic text for an undergraduate course, but students will have to consult other books and publications to find more detailed coverage of particular topics such as laboratory testing, seepage, slope stability and foundation design.

The treatment of soil mechanics and geotechnical engineering in this book is simple, straightforward and largely idealized. I have tried to relate the behaviour of soils and geotechnical structures to everyday experiences, encouraging students to perform simple experiments themselves at home, on holiday and in a basic soil mechanics laboratory. I have described some simple tests which are designed to demonstrate the basic principles rather than generate highly accurate results. Only a few details are given of the apparatus and procedures since engineers should be trained to design and build simple equipment and work out how to make observations and analyse results themselves.

To illustrate the basic nature of soil strength and stiffness I have described the behaviour of soils in oedometer tests and in ideal shear tests in order to separate the effects of normal stress and compression from the effects of shearing and distortion. I have also described the behaviour of soils in triaxial tests, as these are the best tests to evaluate soil parameters. Readers will notice that I have not included data from tests on real soils or case histories of construction performance. This is quite deliberate and is common practice in undergraduate texts on structures, hydraulics, concrete and so on. As the book is intended primarily as an undergraduate teaching text it is kept simple and straightforward. The basic soil mechanics theories have been clearly demonstrated in earlier books from *Critical State Soil Mechanics* by Schofield and Wroth in 1968 to *Soil Behaviour and Critical State Soil Mechanics* by Muir Wood in 1991, and almost everything in this book follows from these well-established theories.

Throughout I have dealt with simple theories and idealizations for soil behaviour. I am very well aware that many natural soils behave in ways that differ from these idealizations and that there are a number of additional factors that may influence the design and analysis of geotechical structures. Nevertheless, I am convinced that for the purposes of teaching the fundamental principles to students it is better to maintain the simplicity of the idealized treatment, provided always that they appreciate that it is idealized. At many points in the text I have indicated where the behaviour of various natural soils may depart significantly from the idealized behaviour. I expect that individual lecturers will bring in other examples of the behaviour of natural soils drawn from their own experiences, but I hope that they would discuss these within the simple framework described in this book.

At the end of most chapters there is a short summary of the main points covered in the chapter and, in most cases, simple worked examples and exercises that illustrate the theories developed in the text. There is also a short selection of books and articles for further reading and a list of specific references quoted in the text.

The courses at City University which form the basis of this book were developed jointly with my colleagues Neil Taylor, Matthew Coop and John Evans and I am grateful to them for their contributions and for their comments and criticisms. I am grateful also for the very detailed comments that I received from many friends and colleagues, including Mark Allman, Eddie Bromhead, Peter Fookes, Charles Hird, Marcus Matthews, Sarah Stallebrass and Giulia Viggiani. The typing was shared between Anne-Christine Delalande and Robert Atkinson.

John Atkinson
City University
London

The SI system of units has been used: the basic units of measurement are:

Length   m
Time   s
Force   N   multiples   kiloNewton $1\,\text{kN} = 10^3\,\text{N}$
                                           megaNewton $1\,\text{MN} = 10^6\,\text{N}$

Some useful derived units are:

Velocity   m/s
Acceleration m/s^2
Stress (pressure) $\text{kN/m}^2 = \text{kiloPascal} = \text{kPa}$
Unit weight kN/m^3

Unit force (1 N) gives unit mass (1 kg) unit acceleration ($1\,\text{m/s}^2$). The acceleration due to the Earth's gravity is $g = 9.81\,\text{m/s}^2$; hence the force due to a mass of 1 kg at rest on Earth is 9.81 N. (*Note*: there are about 10 apples in 1 kg: hence a stationary apple gives rise to a force of about 1 N acting vertically downwards.)

As in most branches of science and engineering, geotechnical engineering uses mathematics and symbols to develop general theories. Because the English alphabet has a limited number of characters use is made of the Greek alphabet:

A	α	alpha
B	β	beta
Γ	γ	gamma
Δ	δ	delta
E	ε	epsilon
Z	ζ	zeta
H	η	eta
Θ	θ	theta
I	ι	iota
K	κ	kappa
Λ	λ	lambda
M	μ	mu
N	ν	nu
Ξ	ξ	xi
O	o	omicron
Π	π	pi
P	ρ	rho
Σ	σ	sigma
T	τ	tau
Y	υ	upsilon
Φ	ϕ	phi
X	χ	chi
Ψ	ψ	psi
Ω	ω	omega

Stress and strain parameters

One-dimensional compression and shear tests:

τ' shear stress
σ' normal stress
γ shear strain
ε_v volumetric strain = normal strain

Triaxial tests:

$q' = (\sigma'_a - \sigma'_r)$	deviatoric stress
$p' = \frac{1}{3}(\sigma'_a + 2\sigma'_r)$	mean normal stress
$\varepsilon_s = \frac{2}{3}(\varepsilon_a - \varepsilon_r)$	shear strain
$\varepsilon_v = \varepsilon_a + 2\varepsilon_r$	volumetric strain

Superscripts for strains

e elastic
p plastic

Subscripts for states

0 initial state (i.e. q'_0, p'_0, v_0)
f critical state (i.e. q'_f, p'_f, v_f)
p peak state (i.e. q'_p, p'_p, v_p)

Subscripts for axes

z, h vertical and horizontal
a, r axial and radial

Normalizing parameters

$\ln p'_c = (\Gamma - v)/\lambda$
$\quad v_\lambda = v + \lambda \ln p'$
$\log \sigma'_c = (e_\Gamma - e)/C_c$
$\quad e_\lambda = e + C_c \log \sigma'$

A	area
A	activity
B	breadth or width
C_c	slope of the normal compression line
C_s	slope of a swelling and recompression line
D	depth
D_r	relative density
E	Young's modulus (E' for effective stress; E_u for undrained loading)
F_s	factor of safety
G_s	specific gravity of soil grains

G	shear modulus (G' for effective stress; G_u for undrained loading)
H	height or thickness
H	maximum drainage path
I_σ	influence coefficient for stress
I_ρ	influence coefficient for settlement
K'	bulk modulus
K_0	coefficient of earth pressure at rest
K_a	coefficient of active earth pressure
K_p	coefficient of passive earth pressure
L	length
LL	liquid limit
LI	liquidity index
N	normal force
N_c, N_γ, N_q	bearing capacity factors
P	potential
P	force on retaining wall
P_a	force due to active pressure
P_p	force due to passive pressure
P_w	force due to free water
Q	flow (volume)
Q	pile load
Q_b	pile base resistance
Q_s	pile shaft resistance
R	radius
S_s	stress state parameter $= p_c'/p'$
S_v	volume state parameter $= v_c - v$
T	shear force
T_v	time factor for one-dimensional consolidation
T_r	time factor for radial consolidation
U	force due to pore pressures
U_t	average degree of consolidation
V	volume
V_w	volume of water
V_s	volume of soil grains
V	velocity (of seepage)
W	work
W	weight
W_w	weight of water
W_s	weight of soil grains
b	thickness or width
c'	cohesion intercept in Mohr–Coulomb failure criterion
c_v	coefficient of consolidation
e	voids ratio
e_0	voids ratio of normally consolidated soil at $p' = 1.0 \, \text{kPa}$
e_κ	voids ratio of overconsolidated soil at $p' = 1.0 \, \text{kPa}$
e_Γ	voids ratio of soil on the critical state line at $p' = 1.0 \, \text{kPa}$
g	shear modulus for states inside the state boundary surface
h_w	height of water in standpipe
i	slope angle

i_c	critical slope angle
i	hydraulic gradient
i_c	critical hydraulic gradient
k	coefficient of permeability
m_v	coefficient of compressibility for one-dimensional compression
q	rate of seepage
q	bearing pressure
q_c	bearing capacity
q_n	net bearing pressure
q_a	allowable bearing pressure
r	radius
r_u	pore pressure coefficient
s	length along a flowline
s_u	undrained strength
t	time
u	pore pressure
u_0	steady state pressure
$\bar{u}$	excess pore pressure
v	specific volume
v_κ	specific volume of overconsolidated soil at $p' = 1.0 \, \text{kPa}$
w	water content
Γ	specific volume of soil on the critical state line at $p' = 1.0 \, \text{kPa}$
Δ	large increment of
M	slope of CSL projected to $q':p'$ plane
N	specific volume of normally consolidated soil at $p' = 1.0 \, \text{kPa}$
$\sum$	sum of
α	adhesion factor for pile friction
γ	unit weight
γ_d	dry unit weight
γ_w	unit weight of water $(= 9.81 \, \text{kN/m}^3)$
δ	small increment of
δ'	angle of friction between structure and soil
η	q'/p'
κ	slope of swelling and recompression line
λ	slope of normal consolidation line and CSL
v	Poisson's ratio (v' for drained loading, $v_u = \frac{1}{2}$ for undrained loading)
ρ	settlement
ρ_c	consolidation settlement
ρ_i	initial settlement
ρ_t	settlement at time t
ρ_∞	final consolidation settlement
ϕ'	angle of friction
ϕ'_a	allowable friction angle
ϕ'_c	critical state friction angle
ϕ'_p	peak friction angle
ψ	angle of dilation

INTRODUCTION TO GEOTECHNICAL ENGINEERING

1.1 WHAT IS GEOTECHNICAL ENGINEERING?

The use of engineering soils and rocks in construction is older than history and no other materials, except timber, were used until about 200 years ago when an iron bridge was built by Abraham Darby in Coalbrookdale. Soils and rocks are still one of the most important construction materials used either in their natural state in foundations or excavations or recompacted in dams and embankments.

Most people have some direct personal experience of soil mechanics and geotechnical engineering. Children at the beach digging holes in the sand, making sand-castles and building dams across streams or, in the kitchen playing with sugar, salt or flour, or in the playroom using plasticine, or in the country losing their boots in the mud are learning about soil mechanics and geotechnical engineering. Soils behave in a variety of different ways. Dry sand will pour like water but it will form a cone, and you can make a sand-castle and measure its compressive strength as you would a concrete cylinder. (The easiest way to measure the strength of a sand-castle is to put it on a balance and press down on it with your hand.) Clay behaves more like plasticine or butter. If the clay has a high water content it squashes like warm butter, but if it has a low water content it is brittle like cold butter and it will fracture and crack if it is compressed. The mechanics that govern the stability of a small excavation or a small slope and the bearing capacity of boots in soft mud are exactly the same as for large excavations and foundations.

Engineering soils are mostly just broken up rock, which is sometimes decomposed into clay, so they are simply collections of particles. In the ground they are usually saturated so the void spaces between the grains are filled with water. Rocks are really strongly cemented soils but they are often cracked and jointed so they are like soil in which the grains fit very closely together. Natural soils and rocks appear in other disciplines such as agriculture and mineral exploitation but in these cases their biological and chemical properties are more important than their mechanical properties. Soils are granular materials and principles of soil mechanics are relevant to storage and transportation of other granular materials such as mineral ores and grain.

Figure 1.1 illustrates a range of geotechnical structures. Except for the foundations, the retaining walls and the tunnel lining all are made from natural geological materials. In slop

Figure 1.1 Examples of geotechnical engineering construction.

and retaining walls the soils apply the loads as well as provide strength and stiffness. Geotechnical engineering is simply the branch of engineering that deals with structures built of, or in, natural soils and rocks. The subject requires knowledge of strength and stiffness of soils and rocks, methods of analyses of structures and hydraulics of groundwater flow.

Use of natural soil and rock makes geotechnical engineering different from many other branches of engineering and more interesting. The distinction is that most engineers can select and specify the materials they use, but geotechnical engineers must use the materials that exist in the ground and they have only very limited possibilities for improving their properties. This means that an essential part of geotechnical engineering is a ground investigation to determine what materials are present and what their properties are. Since soils and rocks were formed by natural geological processes, knowledge of geology is essential for geotechnical engineering.

1.2 PRINCIPLES OF ENGINEERING

Engineers design a very wide variety of systems, machines and structures from car engines to very large bridges. A car engine has many moving parts and a number of mechanisms, such as the pistons, connecting rods and crankshaft or the camshaft and valves, while a bridge should not move very much and it certainly should not form a mechanism. Other branches of engineering are concerned with the production and supply of energy, the manufacture of washing machines and personal computers, the supply, removal and cleaning of water, moving vehicles and goods and so.

Within civil engineering the major divisions are structural (bridges and buildings), hydraulic (moving water) and geotechnical (foundations and excavations). These are all broadly similar in the sense that a material, such as steel, water or soil, in a structure, such as a bridge, river or foundation, is loaded and moves about. The fundamental principles of structural, hydraulic and geotechnical engineering are also broadly similar and follow the same fundamental laws of mechanics. It is a pity that these subjects are often taught separately so that the essential links between them are lost.

In each case materials are used to make systems or structures or machines and engineers use theories and do calculations that demonstrate that these will work properly; bridges must not fall down, slopes or foundations must not fail and nor must they move very much. These theories must say something about the strength, stiffness and flow of the materials and the way the whole structure works. They will deal with ultimate states to demonstrate that the structure does not fall down and they will deal also with working states to show that the movements are acceptable.

Notice that engineers do not built or repair things; they design them and supervise their construction by workers. There is a common popular misconception about the role of engineers. The general public often believes that engineers build things. They do not; engineers design things and workmen build them under the direction of engineers. Engineers are really applied scientists, and very skilled and inventive ones at that.

1.3 FUNDAMENTALS OF MECHANICS

In any body, framework or mechanism changes of loads cause movements; for example a rubber band stretches if you pull it, a tall building sways in the wind and pedalling a bicycle turns the wheels. The basic feature of any system of forces and displacements and stresses and strains are illustrated in Fig. 1.2. Forces give rise to stresses and these must be in equilibrium

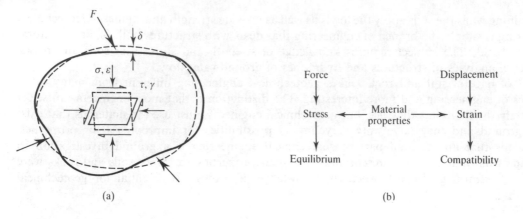

Figure 1.2 Principles of mechanics.

or the body will accelerate. Displacements give rise to strains which must be compatible so the material does not tear or overlap. (Relationships between forces and stresses and between displacements and strains are given in Chapter 2.) These two separate requirements (of equilibrium and compatibility) are quite simple and they apply universally to everything. The relationships between stresses and strains (or between forces and displacements) are governed by the characteristics of the material.

There are a number of branches or subdivisions of mechanics which depend on the material, the type of problem and any assumptions made. Obviously soil mechanics is the mechanics of structures made of soils and there are also rock mechanics for rocks and fluid mechanics for fluids. Some important branches of mechanics are illustrated in Fig. 1.3; all of these are used in soil mechanics and appear later in this book.

Rigid body mechanics deals with mechanisms, such as car engines, in which all the moving parts are assumed to be rigid and do not deform. Structural mechanics is for framed structures where deformations arise largely from bending of beams and columns. Fluid mechanics is concerned with the flow of fluids through pipes and channels or past wings, and there are various branches depending on whether the fluid is compressible or not. Continuum mechanics deals with stresses and strains throughout a deforming body made up of material that is continuous (i.e. it does not have any cracks or joints or identifiable features), while particulate mechanics synthesizes the overall behaviour of a particulate material from the response of the individual grains. You might think that particulate mechanics would be relevant to soils but most of current soil mechanics and geotechnical engineering is continuum mechanics or rigid body mechanics.

1.4 MATERIAL BEHAVIOUR

The link between stresses and strains is governed by the properties of the material. If the material is rigid then strains are zero and movements can only occur if there is a mechanism. Otherwise materials may compress (or swell) or distort, as in Fig. 1.4. Figure 1.4(a) shows a block of material subjected to a confining pressure σ and Fig. 1.4(c) shows a relationship between the pressure and the change of volume; the gradient is the bulk modulus K. The stress can be raised more or less indefinitely and the material continues to compress in a stable manner and does not fail; K continues to increase with stress and strain.

(a) Continuum (b) Particulate

(c) Rigid body (d) Structural

(e) Fluid

Figure 1.3 Branches of mechanics used in geotechnical engineering.

(a)

(b)

$$\varepsilon_v = \frac{\Delta V}{V_0}$$

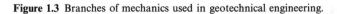

(c) (d)

Figure 1.4 Compression and distortion.

Figure 1.4(b) shows a block of material subjected to shearing stresses τ so that it distorts in shear. Notice that compression in Fig. 1.4(a) involves a change of size while shear distortion involves a change of shape; in a general loading, compression and distortion occur simultaneously. Figure 1.4(d) shows a simple relationship between shear stress and shear strain; the gradient is the shear modulus G and this reduces with stress and strain. The material fails when no more shear stress can be added and then it continues to strain at constant shear stress τ_f; this is the shear strength of the material.

Figure 1.4 illustrates the two most important aspects of material behaviour: stiffness and strength. Stiffness relates changes of stress and changes of strain by

$$K = \frac{d\sigma}{d\varepsilon_v} \qquad G = \frac{d\tau}{d\gamma} \tag{1.1}$$

where $\varepsilon_v = \Delta V / V_0$ is the volumetric strain and γ is the shear strain. The simplest theory for stiffness is the theory of elasticity in which K and G are constants and apply equally to loading and unloading.

Strength is the limiting shear stress that the material can sustain as it suffers large shear strains. The two most common theories for strength are to say that the material is cohesive and the limiting shear stress is a constant for the material given by

$$\tau_f = s \tag{1.2}$$

or to say that the material is frictional so that the strength is proportional to the confining pressure given by

$$\tau_f = \sigma\mu = \sigma \tan \phi \tag{1.3}$$

where μ is a coefficient of friction and ϕ is a friction angle. Later we will find that both of these theories apply to soils, but in different circumstances.

Values for the stiffness parameters K and G and the strength parameters s and μ (or ϕ) will obviously depend on the material, but they may also depend on other things such as temperature and rate of loading. For example, if the strength depends on the rate of strain the material is said to be viscous. The first part of this book, up to Chapter 14, deals largely with the basic theories for the strength and stiffness of soils and other granular materials.

1.5 BASIC CHARACTERISTICS OF SOILS

At first sight soils appear to behave rather strangely. For example, you can pour dry sand like water and you can pour saturated sand under water in the same way, yet you can make sand-castles from partly saturated sand that will support loads like a concrete cylinder. Clays can be squeezed and moulded like plasticine and appear to behave very differently from sands, but very old slopes in clay have angles comparable to those in sands.

The essential features of soil behaviour which we will examine later in this book are as follows:

1. External loads and water pressures interact with each other to produce a stress that is effective in controlling soil behaviour.
2. Soil is compressible; volume changes occur as the grains rearrange themselves and the void space changes.

3. Soil shearing is basically frictional so that strength increases with normal stress, and with depth in the ground. We will find that soil stiffness also increases with normal stress and depth.
4. Combining these basic features of soil behaviour leads to the observation that soil strength and stiffness decrease with increasing water pressure and with increasing water content.
5. Soil compression and distortion are generally not fully recoverable on unloading, so soil is essentially inelastic. This is a consequence of the mechanism of compression by rearrangement of the grains; they do not un-rearrange on unloading.

We will see later that there is no real distinction between sands and clays and that the apparent differences arise from the influence of pore pressures and seepage of water in the void spaces between the grains.

1.6 BASIC FORMS OF GEOTECHNICAL STRUCTURE

The four basic types of geotechnical structure are illustrated in Fig. 1.5; most other cases are variations or combinations of these. Foundations (Fig. 1.5a) transmit loads to the ground and the basic criterion for design is that the settlements should be relatively small. The variables of a foundation are the load F, the size of the base B and the depth D. Foundations may support loads that are relatively small, such as car wheels, or relatively large, such as a power station. Slopes (Fig. 1.5b) may be formed naturally by erosion or built by excavation or filling. The basic variables are the slope angle i and the depth H, and the design requirement is that the slope should not fail by landsliding.

Slopes that are too deep and too steep to stand unsupported can be supported by a retaining wall (Fig. 1.5c). The basic variables are the height of the wall H and its depth of burial D, together with the strength and stiffness of the wall and the forces in any anchors or props. The basic requirements for design are complex and involve overall stability, restriction of ground

(a) Foundation   (b) Slope stability

(c) Retaining wall   (d) Earth-fill dam

Figure 1.5 Geotechnical structures.

movements and the bending and shearing resistance of the wall. In any structure where there are different levels of water, such as in a dam (Fig. 1.5d) or around a pumped well, there will be seepage of water. The seepage causes leakage through a dam and governs the yield of a well and it also governs the variation of pressure in the groundwater.

The structures in Fig. 1.5 clearly should not fail. There are, however, situations where the material must fail; these include ploughing and flow of mineral ore or grain from a storage silo. Solution to problems of this kind can be found using the theories of soil mechanics. Other problems in geotechnical engineering include movement of contaminants from waste repositories and techniques for ground improvement.

1.7 FACTORS OF SAFETY AND LOAD FACTORS

All structural and geotechnical analyses contain uncertainties of one kind or another. These may involve uncertainties in prediction of maximum loads (particularly live loads due to wind, waves and earthquakes) approximations in the theories adopted for material behaviour and structural analysis, and uncertainties in the determination of strength and stiffness parameters. To take account of these approximations and uncertainties it is usual to apply a factor of safety in the design. These factors may be applied as partial factors to reflect the various uncertainties or as a single lumped value.

All applied sciences that analyse and predict natural events involve assumptions, approximations and simplifications because the real world is very complicated. Many people believe that the uncertainties in geotechnical engineering are very large because of the variability of natural soils in the ground and the apparent complexity of theoretical soil mechanics. It is true that geotechnical engineering is less exact than many applications of physics and chemistry, but it is probably less approximate than, say, sociology and economics. You can usually, but not always, improve a theory by making it more complicated and by adding more variables. For example, if material strength and stiffness parameters are allowed to vary with ambient temperature the theories will become more complex but possibly more realistic. In this book I shall be dealing with fairly simple theories of soil mechanics and geotechnical engineering which are suitable for most routine design problems.

Although it is always essential to consider the ultimate collapse states of structures to demonstrate that they will not collapse, the principal design criterion for many structures, particularly foundations, is the need to limit ground movements and settlements. In practice this is often done by applying a factor of safety to the design. In my first job as a young engineer I was involved in the design of a very large earthfill dam where the consequences of collapse would have been catastrophic and would certainly have meant major loss of life: the chief engineer required a factor of safety of about 1.25 against slope failure. In my second job I was asked to design the foundations for a small store shed which was part of a water treatment works: the chief engineer required a factor of safety of 2.5 to 3.

I was puzzled by this inconsistency until I discovered that the large factor required for the foundations of the store shed was not really a factor of safety but was a load factor to limit the settlement. The chief engineer knew that if the collapse load of a foundation was reduced by a factor of 2.5 to 3 the resulting settlements would be small. The point is illustrated in Fig. 1.6 which shows the settlement ρ of a foundation with loading F. In Fig. 1.6(b) there is a collapse load F_c and a load F_r that is about 80 per cent of F_c, corresponding to a factor of safety of about 1.25. There is also an allowable load F_a corresponding to a load factor of about 3, and for this load the settlements are small. In geotechnical engineering it is essential to distinguish between a factor of safety which is intended to account for uncertainties and a load factor which is intended to limit settlements and ground movements.

Figure 1.6 Factor of safety and load factor for a foundation.

1.8 SUMMARY

1. Geotechnical engineering is a branch of engineering and deals with the analysis and design of foundations, slopes and structures made from soils and rocks.
2. The basic theories of mechanics (equilibrium and compatibility) and of material behaviour (stiffness and strength) apply equally in geotechnical engineering.
3. The basic behaviour of soil is influenced both by the loads on the soil grains and the pressures in the water in the void spaces.
4. Soil mechanics describes the relationships between stresses and strains in soils. These will be dealt with in Chapters 8 to 14. We will find that soil behaviour is essentially frictional, compressible and largely inelastic.
5. Methods and theories for analysis and design of geotechnical structures, such as foundations, slopes and retaining walls, and for seepage of groundwater will be covered in Chapters 17 to 23.
6. In geotechnical design safe loads are found by applying factors of safety while ground movements are often restricted by applying a load factor.

FURTHER READING

Chadwick, A. and J. Morfet (1986) *Hydraulics in Civil Engineering*, Allen and Unwin, London.
Gordon, J. E. (1968) *The New Science of Strong Materials*, Penguin, Harmondsworth.
Gordon, J. E. (1978) *Structures*, Penguin, Harmondsworth.
Montague, P. and R. Taylor (1989) *Structural Engineering*, McGraw-Hill, London.
Palmer, A. C. (1976) *Structural Mechanics*, Oxford University Press.
Spencer, A. J. M. (1980) *Continuum Mechanics*, Longman, London.
Upton, N. (1975) *An Illustrated History of Civil Engineering*, Heinemann, London.

CHAPTER TWO

BASIC MECHANICS

This chapter, and the following one, cover the basic methods for the analysis of stress and strain using the Mohr circle constructions and the general features of material behaviour. These techniques are essential for understanding soil behaviour and for analysing soil structures and will be used extensively throughout the book. The topics should be covered in other courses on strength of materials, but here they are put into the context of soil mechanics. Readers are advised to skim through these two chapters and come back to them to work through the details as necessary.

2.1 INTRODUCTION

Mechanics is the study of forces and displacements, or stresses and strains, and there are a number of branches of mechanics associated with particular materials or with particular applications. The fundamental principles of mechanics are simply the application of equilibrium and compatibility. For any body that is not accelerating the forces and moments must be in equilibrium: this is simply Newton's first law. For any body, or system of bodies, that is distorting or moving about the strains and displacements must be compatible. This means that material does not vanish and gaps do not appear; this is simply common sense. What we can do is to analyse states of stress (or strain) so that we can calculate the stresses (or strains) on any plane at a point from the stresses (or strains) on any other pair of planes.

2.2 STRESSES AND STRAINS

I shall assume that readers have been introduced to the basic ideas of stress and strain in other courses. A stress is basically an intensity of loading given by a force acting on a unit area while a strain is basically an intensity of deformation given by a displacement over a unit gauge length. In geotechnical engineering there are two minor differences from the definitions of stress and strain usually adopted for metals and concrete, and these account for the particulate nature of soils. Firstly, the unit area or gauge length must be large enough to include a representative

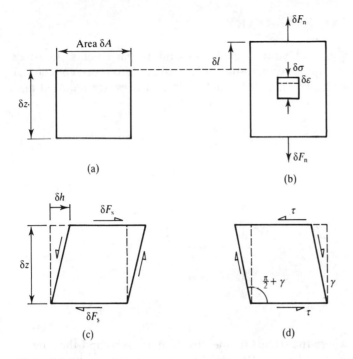

Figure 2.1 Stress and strain.

number of soil grains and, secondly, because uncemented soils cannot sustain tensile stresses compressive stresses are positive.

Figure 2.1 shows stresses and strains in a cube of soil subjected to normal and shear forces. The changes of normal stress $\delta\sigma$ and normal strain $\delta\varepsilon$ due to a change of normal load δF_n are given by

$$\delta\sigma = -\frac{\delta F_n}{\delta A} \tag{2.1}$$

$$\delta\varepsilon = -\frac{\delta l}{\delta z} \tag{2.2}$$

(Notice that negative signs have been added so that compressive stresses and strains are positive quantities.) The changes of shear stress $\delta\tau$ and shear strain $\delta\gamma$ due to a change of load δF_s are given by

$$\delta\tau = -\frac{\delta F_s}{\delta A} \tag{2.3}$$

$$\delta\gamma = -\frac{\delta h}{\delta z} \tag{2.4}$$

(Notice that negative signs have been added so that positive shear stresses and shear strains are associated with increases in the angles in the positive quadrants of the element as shown.)

2.3 PLANE STRAIN AND AXIAL SYMMETRY

In general we should consider stresses and strains, or forces and displacements, in three dimensions, but then the algebra becomes quite complicated and it is difficult to represent general states on flat paper. There are, however, two cases for which only two axes are required and these are illustrated in Fig. 2.2.

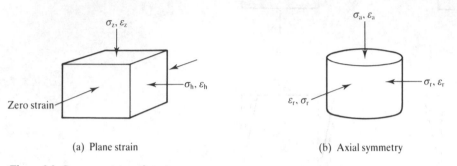

(a) Plane strain (b) Axial symmetry

Figure 2.2 Common states of stress.

Figure 2.2(a) shows plane strain where the strains in one direction are zero and the stresses and strains are vertical (σ_z, ε_z) or horizontal (σ_h, ε_h). (It would be best to use v as the subscript for vertical stress and strain but we will need to keep the subscript v for volumes and volumetric strains.) This corresponds to conditions in the ground below a long structure, such as an embankment or wall or a strip foundation. Figure 2.2(b) shows axial symmetry where the radial stresses and strains (σ_r, ε_r) are equal and the other stresses and strains (σ_a, ε_a) are axial. This corresponds to conditions in the ground below a circular foundation or a circular excavation. Throughout this book I will consider only plane strain and axial symmetry and I will use the axes z, h (vertical and horizontal) for plane strain and the axes a, r (axial and radial) for axial symmetry.

2.4 RIGID BODY MECHANICS

When soils fail they often develop distinct slip surfaces; on a geological scale these appear as faults. Slip surfaces divide soil into blocks and the strains within each block may be neglected compared with the relative movements between blocks, so the principles of rigid body mechanics are applicable for failure of slopes and foundations. To demonstrate this take a block of butter from the fridge (remove the paper first), put it on end and press down hard on it (i.e. do a compression test). The block will almost certainly develop a slip plane at about 45° to the horizontal. (Does the temperature of the butter make any difference?)

Equilibrium is examined by resolution of forces in two directions (together with moments about one axis) and this is done most simply by construction of a polygon of forces: if the polygon of forces closes then the system of forces is in equilibrium. Figure 2.3(a) shows a set of forces acting on a triangular block. We will see later than this represents the conditions in soil behind a retaining wall at the point of failure. Figure 2.3(b) shows the corresponding polygon of forces where each line is in the same direction as the corresponding force and the length is proportional to the magnitude of the force. The forces are in equilibrium because the polygon of forces is closed. You will study conditions for equilibrium of forces in other courses on mechanics or strength of materials and this is just the same.

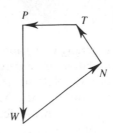

(a) Forces on a body

(b) Polygon of forces

Figure 2.3 Conditions of equilibrium.

(a) Mechanism of displacement

(b) Displacement diagram

Figure 2.4 Conditions of compatibility.

Compatibility of displacement is examined most conveniently by construction of a displacement diagram (also known as a hodograph). Figure 2.4(a) illustrates two triangular blocks moving as illustrated by the arrows; we will see later that this could represent the displacement of soil below a foundation at the point of failure. Each block is given an identifying letter and O represents stationary material. In Fig. 2.4(b) each arrow represents the direction and magnitude of the displacement of one of the rigid blocks and the displacement diagram closes. (Note that the displacement diagram in Fig. 2.4b is different from the force polygon in Fig. 2.3b in that the arrows are not all in the same direction.) The letters on the arrows represent the relative displacements, thus oa is the displacement of A with respect to O and ab is the displacement of B with respect to A.

The relative movements of rigid bodies in mechanisms like that shown in Fig. 2.4 can be examined by making simple models from stiff card. (From a flat sheet of card cut a triangular recess, cut two triangular shapes like A and B and demonstrate that you have a compatible mechanism. To get them to move it is necessary to drill small holes at the corners.)

2.5 ANALYSIS OF STRESS

Within a loaded body the stresses generally vary from point to point so, for example, the stresses below the edge and centre of a foundation are different. At any point the stresses are different on different planes and it is necessary to relate the stresses on the different planes.

The simplest form of analysis is through the Mohr circle construction which is covered in courses on strength of materials. The only difference for soil mechanics is that the sign convention is changed so that compressive stresses and counter-clockwise shear stresses are positive.

Figure 2.5(a) shows principal stresses σ_z and σ_h on the faces of an element of soil and Fig. 2.5(b) shows the corresponding Mohr circles of stress. The pole P of the Mohr circle is defined so that a line from P to σ_z gives the direction of the plane on which σ_z acts. In Fig. 2.5(a) there is an element rotated to an angle θ as shown and the stresses (τ_n, σ_n and τ_m, σ_m) on the faces of this element are at N and M in Fig. 2.5(b). From the geometry of the Mohr circle the angle 2θ subtended at the centre by the point representing the major principle plane and the point N is twice the angle between the planes on which these stresses act. From the geometry of the figure, $\tau_n = \tau_m$. Using Fig. 2.5(b) it is possible to calculate τ_n, σ_n and τ_m, σ_m from σ_z and σ_h or vice versa, and in order to construct the Mohr circle it is necessary to know the stresses on two (preferably orthogonal) planes.

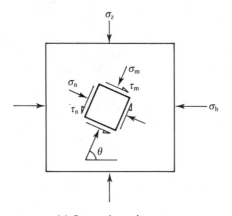

(a) Stresses in an element

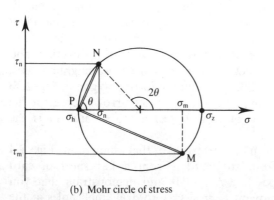

(b) Mohr circle of stress

Figure 2.5 Analysis of stress.

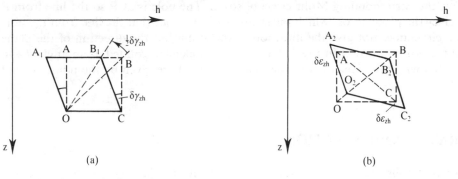

Figure 2.6 Shear strains in an element.

2.6 ANALYSIS OF STRAIN

Analysis of strains at a point using the Mohr circle of strain is similar to that for stress, but there are a few points to note about strains. Firstly, while it is possible to talk about a state of stress with respect to zero stress (taken as atmospheric pressure), there is no absolute zero for strain so we have to talk about changes, or increments, of strain. These may be small increments (denoted by $\delta\varepsilon$) or large increments (denoted by $\Delta\varepsilon$) and generally they occur as a result of corresponding large or small increments of stress. Secondly, while stresses in soils are almost always positive (particulate materials cannot usually sustain tensile stresses unless the grains are attached to one another), strains may be positive (compressive) or negative (tensile) and in an increment of strain there will usually be compressive and tensile strains in different directions. Thirdly, we must be careful to distinguish between pure shear strains and engineers' shear strains $\delta\gamma$ and take account of any displacements of the centre of area of distorted elements.

Figure 2.6(a) shows an element OABC strained by $\delta\gamma_{zh}$ to a new shape OA_1B_1C. It can be seen that the diagonal OB has rotated to OB_1 through $\frac{1}{2}\delta\gamma_{zh}$. Figure 2.6(b) shows the strained element rotated and translated to $O_2A_2B_2C_2$ so that the centre and the diagonals coincide and the edges have now all strained through the same angle $\delta\varepsilon_{zh} = \delta\varepsilon_{hz} = \frac{1}{2}\delta\gamma_{zh}$.

Figure 2.7(a) shows a plane element with principal strains $\delta\varepsilon_z$ and $\delta\varepsilon_h$ (which is negative)

(a) Strains in an element

(b) Mohr circle of strain

Figure 2.7 Analysis of strain.

while Fig. 2.7(b) is the corresponding Mohr circle of strain. The pole is at P so the line from P to the point $\delta\varepsilon_z$ gives the plane across which the strain is $\delta\varepsilon_z$. (Notice that the line from the pole to a point on the circle does not give the direction of the strain but the direction of the plane perpendicular to the normal strain.) In Fig. 2.7(a) there is an element rotated at an angle θ and the strains associated with this element ($\frac{1}{2}\delta\gamma_n$, $\delta\varepsilon_n$ and $\frac{1}{2}\delta\gamma_m$, $\delta\varepsilon_m$) are given by the points N and M as shown.

2.7 STRESS RATIO AND DILATION

We will see later that soils are frictional materials, which means that their strength (i.e. the maximum shear stress they can sustain) increases with normal stress and so the stress ratio τ/σ is more important than the shear stress alone. Figure 2.8(a) shows a stressed element and Fig. 2.8(b) is the corresponding Mohr circle of stress with the pole at P. There are two lines ON which are tangents to the Mohr circle and these define the points on which the stress ratio is given by

$$\frac{\tau}{\sigma} = \tan\phi_{mob} \tag{2.5}$$

where ϕ_{mob} is the mobilized angle of shearing resistance. From the geometry of Fig. 2.8(b) $t = \frac{1}{2}(\sigma_z - \sigma_h)$ and $s = \frac{1}{2}(\sigma_z + \sigma_h)$ and

$$\frac{t}{s} = \sin\phi_{mob} = \frac{\sigma_z - \sigma_h}{\sigma_z + \sigma_h} \tag{2.6}$$

or

$$\frac{\sigma_z}{\sigma_h} = \frac{1 + \sin\phi_{mob}}{1 - \sin\phi_{mob}} = \tan^2(45° + \tfrac{1}{2}\phi_{mob}) \tag{2.7}$$

The planes, shown by double lines, on which this stress ratio occurs are at angles α and β as shown and, from the geometry of the figure,

$$\alpha = \beta = 45° + \tfrac{1}{2}\phi_{mob} \tag{2.8}$$

For frictional materials these correspond to the planes on which the most critical conditions occur and they should be the planes on which failure will occur.

When the major and minor principal strains have opposite signs the origin of the axes is inside the Mohr circle, as shown in Fig. 2.9(b). There are two planes, shown by broken lines in Fig. 2.9(b), across which the normal strains are zero, and so there are two directions, shown by double lines, at angles α and β along which the strains are zero as shown in Fig. 2.9(a). These planes are defined by an angle of dilation ψ. From Fig. 2.9(b), the lengths $v = \frac{1}{2}(\delta\varepsilon_z + \delta\varepsilon_h)$ and $g = \frac{1}{2}(\delta\varepsilon_z - \delta\varepsilon_h)$, and if the volumetric strain is $\delta\varepsilon_v = \delta\varepsilon_z + \delta\varepsilon_h$ then the angle of dilation is given by

$$\tan\psi = -\frac{\delta\varepsilon_v}{\delta\gamma} \tag{2.9}$$

or

$$\sin\psi = -\frac{\delta\varepsilon_z + \delta\varepsilon_h}{\delta\varepsilon_z - \delta\varepsilon_h} \tag{2.10}$$

where $\delta\gamma$ is the increment of shear strain across the plane. (The negative signs are required in

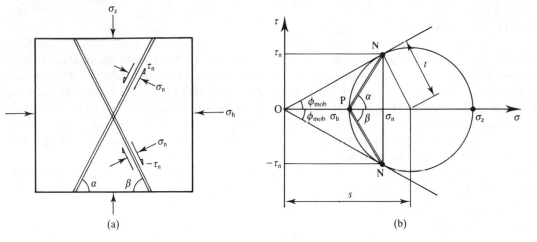

Figure 2.8 Limiting stress ratio and angle of shearing resistance.

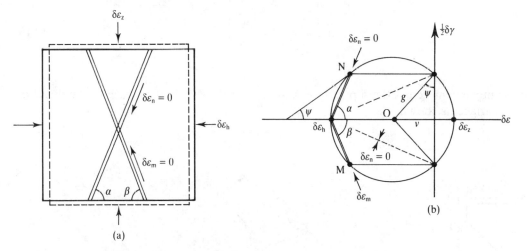

Figure 2.9 Angle of dilation and zero extension lines.

Eqs (2.9) and (2.10) so positive angles ψ are associated with dilation or negative volumetric strains.) From the geometry of the figure,

$$\alpha = \beta = 45° + \tfrac{1}{2}\psi \tag{2.11}$$

Comparing Figs 2.8 and 2.9, the angle of dilation ψ describing the strain ratio $\delta\varepsilon_v/\delta\gamma$ has similar properties to the angle of shearing resistance ϕ_{mob} which describes the stress ratio τ/σ.

You can visualize how materials strain by drawing a circle with a felt-tipped pen on a sheet of thin rubber and stretching it. The circle will distort into an ellipse and its area may increase. You can probably see that there are two diameters of the original circle that remain the same length and these correspond to the directions of zero strain.

2.8 SLIP SURFACES

Figure 2.9 represents homogeneous straining where there are no discontinuities, or slip surfaces, like those that appeared during rigid body deformation of cold butter and the double lines show the directions of zero strain. Figure 2.10(a) shows material that is deforming by intense shearing in a very thin zone AB and Fig. 2.10(b) shows a detail of the slip zone. This thin zone of shearing material has a small but finite thickness which is usually too small to see; in soils it is probably of the order of ten grains thick. Shear zones usually appear to have no thickness and so they are called slip planes or slip surfaces.

Since the length of AB in Fig. 2.10(a) remains constant, because the material on either side is rigid, it is a zero extension line and its direction is given by $\alpha = 45° + \frac{1}{2}\psi$, as in Fig. 2.9. From Fig. 2.10(b),

$$\delta\gamma = \frac{\delta h}{H_0} \qquad \delta\varepsilon_v = \frac{\delta v}{H_0} \tag{2.12}$$

$$\tan\psi = \frac{\delta\varepsilon_v}{\delta\gamma} = \frac{\delta v}{\delta h} \tag{2.13}$$

so that the movement across the slip surface $A \to A_1$ and $B \to B_1$ is at angle ψ to the direction of the slip surface as shown.

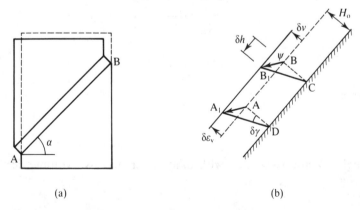

Figure 2.10 Discontinuous slipping and slip surfaces.

2.9 SUMMARY

1. Forces and stresses in any body of material must be in equilibrium: this means that the polygon of forces acting on the body, or on an element inside the body, closes.

2. Strains and displacements in any distorting body must be compatible: this means that the material does not tear or overlap and the displacement diagram (or hodograph) closes.
3. States of stress or strain at a point can be analysed using the Mohr circle construction so that the stresses or strains on any plane can be calculated from the geometry of the circle.
4. If slip surfaces develop, their directions correspond to the directions of zero extension lines and the relative movement across a slip surface is at an angle ψ to its direction.

WORKED EXAMPLES

Example 2.1: Equilibrium of forces using a force polygon Figure 2.11(a) shows forces acting on a rigid triangular block of soil with a slip surface; two of the forces are known to be $W = 160$ kN and $T = 60$ kN. Figure 2.11(b) shows the corresponding polygon of forces. Scaling from the diagram, or by calculation, $P = 75$ kN.

(a)   (b)

Figure 2.11

Example 2.2: Compatible displacements using a displacement diagram Figure 2.12(a) shows two rigid blocks separated by slip surfaces where all the angles are 45° or 90°; the left-hand block moves with a vertical component of displacement 1 mm as shown. Figure 2.12(b) shows the corresponding displacement diagram. Scaling from the diagram, or by calculation, $\delta h = 2$ mm.

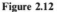

(a)   (b)

Figure 2.12

Example 2.3: Stress analysis using a Mohr circle of stress Figure 2.13(a) shows an element of soil behind a retaining wall; the effective vertical and horizontal stresses are $\sigma_v = 300$ kPa and $\sigma_h = 100$ kPa and these are principal stresses. Figure 2.13(b) shows the Mohr circle of stress. Scaling from the diagram, $\phi_{mob} = 30°$, the angles of the critical planes are $\alpha = \beta = 60°$ and the stresses on these planes are $\sigma = 150$ kPa and $\tau = \pm 87$ kPa.

(a)   (b)

Figure 2.13

Example 2.4: Analysis of strain using a Mohr circle of strain Figure 2.14(a) illustrates an increment of displacement of a retaining wall. The strains in an element of soil behind the wall are $\delta\varepsilon_z = 0.10$ per cent and $\delta\varepsilon_h = -0.20$ per cent and these are principal strains. Figure 2.14(b) shows the Mohr circle for the increment of strain. Scaling from the diagram, the angle of dilation is $\psi = 20°$. The zero extension lines are at angles $\alpha = \beta = 55°$ and the shear strains across zero extension lines are given by $\frac{1}{2}\delta\gamma = \pm 0.14$ per cent.

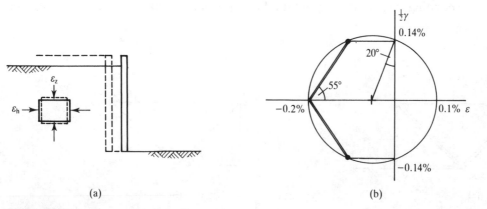

(a)   (b)

Figure 2.14

FURTHER READING

Atkinson, J. H. (1981) *Foundations and Slopes*, McGraw-Hill, London.
Case, J. and A. H. Chilver (1971) *Strength of Materials and Structures*, Edward Arnold, London.

CHAPTER THREE

ESSENTIALS OF MATERIAL BEHAVIOUR

Before reading this chapter, read the note at the beginning of Chapter 2.

3.1 STRESS–STRAIN BEHAVIOUR, STIFFNESS AND STRENGTH

Chapter 2 considered the states of stress and strain at a point in loaded and deforming material. The analyses that were developed for stresses and strains, using Mohr circles, are not dependent on the material in question and they are equally applicable for steel, concrete or soil. In order to analyse any kind of structure, or any kind of solid or fluid continuum, it is necessary to have relationships between stresses and strains. These are called constitutive relationships and they take a number of different forms depending on the nature of the material and on the loading.

Figure 3.1 shows an idealized relationship between stress and strain and it is similar to the stress–strain curves for common engineering materials like metals, plastics, ceramics and engineering soils. For soils and other granular materials, it is necessary to deal with something called effective stress to take account of pore pressures in the fluid in the voids between the grains. (In simple terms effective stresses can be thought of as the stresses effective in the soil grains.) Effective stress will be covered in Chapter 6 where it will be shown that all soil

Figure 3.1 A typical stress–strain curve for soil.

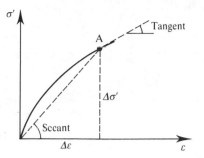

Figure 3.2 Tangent and secant stiffness moduli.

behaviour including stiffness and strength, is governed by an effective stress which is denoted by a prime (as in σ'). For correctness I will use effective stresses from now on.

Stiffness is the gradient of the stress–strain line. If this is linear the gradient is easy to determine but, if it is curved, the stiffness at a point such as A may be quoted as a tangent or as a secant, as shown in Fig. 3.2 and given by

$$\text{tangent stiffness} = \frac{d\sigma'}{d\varepsilon} \tag{3.1}$$

$$\text{secant stiffness} = \frac{\Delta\sigma'}{\Delta\varepsilon} \tag{3.2}$$

It is the stiffness of a material that largely determines the strains and displacements in structures, or in the ground, as they are loaded or unloaded. Another term often used in soil mechanics to describe the relationship between stress and strain is 'compressibility', but this is basically the reciprocal of stiffness. Often there is a marked change in the gradient of a stress–strain curve at a yield point, as shown in Fig. 3.1. This is associated with a fundamental change in behaviour often from elastic and recoverable straining to inelastic and irrecoverable straining.

In simple terms the strength of a material is the largest stress that the material can sustain and it is this which governs the stability or collapse of structures.

Stiffness and strength are quite different things: one governs displacements at working load and the other governs the maximum loads that a structure can sustain. Materials may be stiff (i.e. have high stiffness) or soft and they may be strong or weak and they may have any reasonable combination of stiffness and strength. Steel is stiff and strong while margarine is soft and weak; blackboard chalk is relatively stiff and weak while rubber is relatively soft and strong.

3.2 CHOICE OF PARAMETERS FOR STRESS AND STRAIN

Figure 3.1 shows the characteristics of material behaviour, but axes of stress and strain are not carefully defined. The choice of axes will depend on the tests carried out to examine the material behaviour and the parameters required. For metals that are essentially elastic and then plastic the parameters required are Young's modulus E, Poisson's ratio v and the yield and ultimate stresses, which can be obtained from a simple uniaxial extension test. For concrete, the required parameters can be obtained from a uniaxial compression test. For soils, volume changes that occur during compression and shearing are very important and to describe soil behaviour we must examine separately shearing and volumetric strains and responses to shearing and normal loading and unloading.

(a) Triaxial test (b) Shear test **Figure 3.3** Common soil tests.

(a)

(b)

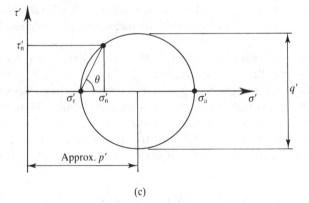

(c)

Figure 3.4 States of stress in triaxial and shear tests.

The two tests commonly used in soil mechanics are the triaxial test and the shear test illustrated in Fig. 3.3. These will be considered in more detail in Chapter 7. The relationships between the stresses in the two tests can be obtained from the Mohr circle construction, as shown in Fig. 3.4. This illustrates that, within the triaxial specimen with stresses (σ'_a, σ'_r) there are elements with stresses (τ'_n, σ'_n) like those in a shear specimen and vice versa.

In the shear test the sample could be loaded or unloaded with zero shear stress and it would compress or swell with normal strains ε_n. Alternatively, it could be sheared to the left or to the right and there would be shear strains. We can define a shear modulus G' and a compression modulus M' as

$$G' = \frac{d\tau'}{d\gamma} \tag{3.3}$$

$$M' = \frac{d\sigma_n'}{d\varepsilon_n} \tag{3.4}$$

(These are written as tangent moduli but they could equally well be secants.) For triaxial tests we need to define parameters for shear and normal stress and strain which give equivalent stiffness moduli. The most convenient parameters are

$$q' = \sigma_a' - \sigma_r' \tag{3.5}$$

$$p' = \tfrac{1}{3}(\sigma_a' + 2\sigma_r') \tag{3.6}$$

$$\varepsilon_s = \tfrac{2}{3}(\varepsilon_a - \varepsilon_r) \tag{3.7}$$

$$\varepsilon_v = \varepsilon_a + 2\varepsilon_r \tag{3.8}$$

and then the shear modulus G' and the bulk modulus K' are

$$3G' = \frac{dq'}{d\varepsilon_s} \tag{3.9}$$

$$K' = \frac{dp'}{d\varepsilon_v} \tag{3.10}$$

The parameter q' is the diameter of the Mohr circle and it is a measure of the maximum shear stress. The parameter p' is the average stress and it is approximately equal to the distance of the centre of the circle from the origin, as shown in Fig. 3.4. The parameter ε_v is simply the volumetric strain and ε_s is equivalent to the shear strain. Comparing Eqs (3.3) and (3.9), both give the same shear modulus G' and, comparing Eqs (3.4) and (3.10), the compression modulus M' (which corresponds to one-dimensional straining) is approximately equal to the bulk modulus K'.

A very helpful distinction to notice is that the volumetric strain ε_v describes the change in size of an element while the shear strain ε_s describes the change in its shape. The value $\tfrac{2}{3}$ in Eq. (3.7) is required for consistency. During an increment of straining the work done per unit volume of soil δW must be invariant (i.e. independent of the choice of parameters). In terms of axial and radial stresses and strains we have

$$\delta W = \sigma_a' \varepsilon_a + 2\sigma_r' \delta\varepsilon_r \tag{3.11}$$

and we also have

$$\delta W = q'\delta\varepsilon_s + p'\delta\varepsilon_v \tag{3.12}$$

You should substitute Eqs (3.5) to (3.9) into Eq. (3.12) and demonstrate that this reduces to Eq. (3.11). When considering the behaviour of soils in triaxial tests I will generally use the parameters q', p', ε_s and ε_v and for shear tests I will generally use the parameters τ_n', σ_n', γ and ε_n.

3.3 CONSTITUTIVE EQUATIONS

During a general loading in the ground both shear and normal stresses are likely to change simultaneously so there will be shearing and volumetric straining together. For soils it turns out that shearing and volumetric effects are coupled so that shearing stresses cause volumetric strains and normal stresses cause shear strains. This is quite surprising and we will see later how the particulate nature of soils gives rise to shear and volumetric coupling.

A simple constitutive equation relating shearing and volumetric stress–strain behaviour can be written as

$$\begin{Bmatrix} \delta q' \\ \delta p' \end{Bmatrix} = \begin{bmatrix} S_{11} & S_{12} \\ S_{21} & S_{22} \end{bmatrix} \begin{Bmatrix} \delta \varepsilon_s \\ \delta \varepsilon_v \end{Bmatrix} \tag{3.13}$$

where $[S]$ is a stiffness matrix containing stiffness moduli. The components of $[S]$ are

$$S_{11} = \frac{\partial q'}{\partial \varepsilon_s} = 3G' \tag{3.14}$$

$$S_{22} = \frac{\partial p'}{\partial \varepsilon_v} = K' \tag{3.15}$$

$$S_{12} = \frac{\partial q'}{\partial \varepsilon_v} = J'_1 \tag{3.16}$$

$$S_{21} = \frac{\partial p'}{\partial \varepsilon_s} = J'_2 \tag{3.17}$$

For materials that are isotropic and elastic and perfectly plastic (see Sec. 3.6), $J'_1 = J'_2$ and the stiffness matrix is symmetric, while for materials that are isotropic and elastic, $J'_1 = J'_2 = 0$ (see Sec. 3.5) so that shearing and volumetric effects are decoupled. Alternatively, a constitutive equation can be written as

$$\begin{Bmatrix} \delta \varepsilon_s \\ \delta \varepsilon_v \end{Bmatrix} = \begin{bmatrix} C_{11} & C_{12} \\ C_{21} & C_{22} \end{bmatrix} \begin{Bmatrix} \delta q' \\ \delta p' \end{Bmatrix} \tag{3.18}$$

where $[C]$ is a compliance matrix containing compliance parameters. Comparing Eqs (3.13) and (3.18), $[C]$ is the inverse of $[S]$ and, in general, there are no simple relationships between the stiffness parameters in $[S]$ and the compliance parameters in $[C]$. However, for materials that are isotropic and elastic, shear and volumetric effects are decoupled so that $C_{12} = C_{21} = 0$ and in this case $C_{11} = 1/S_{11} = 1/3G'$ and $C_{22} = 1/S_{22} = 1/K'$.

Since the stress–strain behaviour of soil is largely non-linear the stiffness and compliance parameters in Eqs (3.13) and (3.18) will not be constants, but will vary with strain. They also depend on the current stresses and on the history of loading and unloading.

3.4 STRENGTH

The strength of a material describes the ultimate state of stress that it can sustain before it fails. (For soils that can suffer very large strains we will have to define failure very carefully, but this will be considered in detail later.) People talk about tensile strength, compressive strength, shear strength, and so on, as though they were all different, but these should really all be related to some fundamental characteristic strength.

The link between these different strengths is the maximum shear stress, or the size of the largest Mohr circle that the material can sustain. Figure 3.5(a) and (b) shows uniaxial tensile and compression tests and the corresponding Mohr circle of stress; the test samples fail when the Mohr circle reaches the limiting size given by the radius τ'_f. Figure 3.5(c) shows a vertical cut and the shear and normal effect stresses on some inclined plane are τ'_n and σ'_n; failure will occur when the Mohr circle reaches its limiting size.

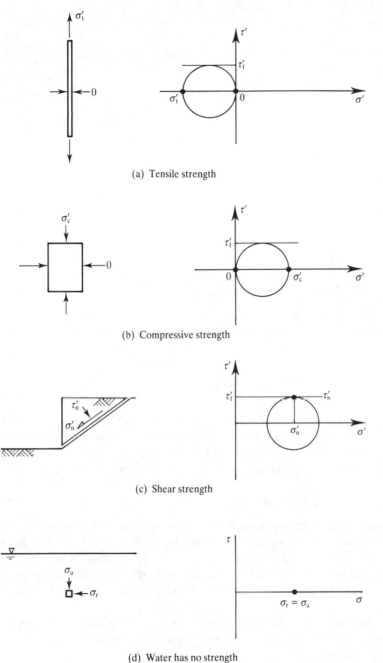

(a) Tensile strength

(b) Compressive strength

(c) Shear strength

(d) Water has no strength

Figure 3.5 Strength of materials.

Thus we can say that materials that have strength can sustain shear stresses and the strength is the maximum shear stress that can be sustained. Only materials with strength can have slopes because shear stresses are required to maintain a slope. A material that cannot sustain a slope, like stationary water, has no strength and the Mohr circle reduces to a point as shown in Fig. 3.5(d).

For soils there are two principal criteria of failure. The first, illustrated in Fig. 3.6(a), is called the Tresca criterion and it says that the material will fail when the Mohr circle of stress

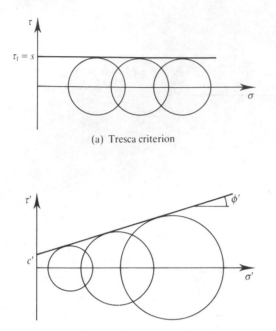

(a) Tresca criterion

(b) Mohr–Coulomb criterion

Figure 3.6 Failure criteria for structural materials.

touches an envelope given by

$$\tau_f = s \tag{3.19}$$

where s is known as the shear strength of the material. (Notice that Eq. (3.19) is written in total, not effective stresses; the reasons for this will be discussed in Chapter 9.) The second, illustrated in Fig. 3.6(b), is called the Mohr–Coulomb criterion and it says that the strength increases linearly with increasing normal effective stress and the material will fail when the Mohr circle touches an envelope given by

$$\tau'_f = c' + \sigma' \tan \phi' \tag{3.20}$$

where ϕ' is the angle of friction and c' is called the cohesion intercept. We will find that $c' = 0$ for the majority of soils so they have no strength when σ' is zero, which is why we can pour dry sand from a jug like water. The poured sand will, however, form a cone, unlike water, showing that it has shear stresses because, inside the cone, the normal stresses are not zero. It turns out that the slope angle i is equal to the friction angle ϕ', so pouring dry soil into a cone is a good way to measure ϕ'.

Dry sugar is a frictional material and its strength is given by Eq. (3.20), while the strength of butter is given by Eq. (3.19) (although the value of s will depend on the temperature of the butter). What about dry flour however?

3.5 ELASTICITY

Materials that are elastic are conservative so that all of the work done by the external stresses during an increment of deformation is stored and is recovered on unloading: this means that all of the strains that occur during an increment of loading are recovered if the increment is removed. An important feature of isotropic and elastic materials is that shear and volumetric effects are decoupled so that the stiffness parameters J'_1 and J'_2 are both zero and Eq. (3.13)

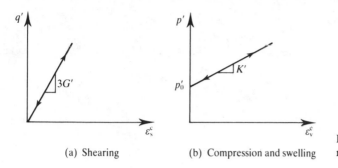

(a) Shearing · (b) Compression and swelling

Figure 3.7 Behaviour of ideal linear elastic material.

becomes

$$\begin{Bmatrix} \delta q' \\ \delta p' \end{Bmatrix} = \begin{bmatrix} 3G' & 0 \\ 0 & K' \end{bmatrix} \begin{Bmatrix} \delta \varepsilon_s^e \\ \delta \varepsilon_v^e \end{Bmatrix} \tag{3.21}$$

(where the superscripts e denote elastic strains) and the complete behaviour is as shown in Fig. 3.7. For materials that are elastic but anisotropic the coupling moduli J_1' and J_2' are equal, so that the matrix in Eq. (3.13) is symmetric about the leading diagonal. Elastic materials can be non-linear, in which case all of the elastic moduli vary with changing stress or strain. (Stretching and relaxing a rubber band is an example of non-linear and recoverable elastic behaviour.)

The more usual elastic parameters are Young's modulus E' and Poisson's ratio v'. These are obtained directly from the results of uniaxial compression (or extension) tests with the radial stress held constant (or zero), and are given by

$$E' = \frac{d\sigma_a'}{d\varepsilon_a^e} \tag{3.22}$$

$$v' = -\frac{d\varepsilon_r^e}{d\varepsilon_a^e} \tag{3.23}$$

Most texts on the strength of materials give the basic relationships among the various elastic parameters and, for isotropic materials, these are

$$G' = \frac{E'}{2(1 + v')} \tag{3.24}$$

$$K' = \frac{E'}{3(1 - 2v')} \tag{3.25}$$

In soil mechanics the shear and bulk moduli, G' and K', are preferable to Young's modulus E' and Poisson's ratio v' because it is important to consider shearing or change of shape separately, or decoupled, from compression or change of size.

3.6 PERFECT PLASTICITY

When the loading has passed the yield point in Fig. 3.1 simultaneous elastic and plastic strains occur and the stiffness decreases. During an increment of plastic deformation the work done is dissipated and so plastic strains are not recovered on unloading. (Bending a paper clip so it remains permanently out of shape is an example of plastic deformation.)

At the ultimate state there are no further changes of stress (because the stress–strain curve

Figure 3.8 Behaviour of ideal perfectly plastic materials.

is horizontal) and so all the strains at failure are irrecoverable. The plastic strains at failure in Fig. 3.1 are indeterminate; they can go on more or less for ever and so we can talk about plastic flow. Although it is impossible to determine the magnitudes of the plastic strains at failure, it is possible to say something about the relative rates of different strains such as shear and volumetric strains.

Figure 3.8(a) illustrates an element of material loaded to failure with different combinations of some arbitrary stresses, σ'_x and σ'_y. The combinations of stress that cause failure and plastic flow are illustrated in Fig. 3.8(b) and are represented by a failure envelope. At any point on the envelope the vector of the failure stress is σ'_f and Fig. 3.8(c) shows the corresponding plastic strains. Since the stresses remain constant the strains accumulate with time and so the origin is arbitrary. The direction of the vector of an increment of the plastic straining is given by $\delta\varepsilon^p_x/\delta\varepsilon^p_y$. The relationship between the failure envelope and the direction of the vector of plastic strain is called a flow rule.

Figure 3.8(d) contains the same information as Fig. 3.8(b) and (c) with the axes superimposed and the origin for plastic strains placed at the end of the appropriate vector of failure stress. For a perfectly plastic material the vector of plastic strain is normal to the failure envelope, and this is known as the normality condition of perfect plasticity.

Another common way of describing the flow rule for plastic straining is to define a plastic potential envelope that is orthogonal to all the vectors of plastic straining, as shown in Fig. 3.9.

(a) (b) **Figure 3.9** Plastic potential.

Figure 3.10 Vector of plastic straining.

Then the material is perfectly plastic if the plastic potential is the same as the failure envelope. This is called an associated flow rule as the plastic potential is associated with the failure envelope. Of course the normality condition and an associated flow rule are different ways of saying the same thing.

An important feature of plastic straining is that the strains depend on the state of stress and do not depend on the small change of stress that causes the failure. This is in contrast to elastic straining where the strains depend on the increments of stress as given by Eq. (3.21). Figure 3.10 shows two different loadings B → A and C → A, both of which cause failure at A. The plastic strains are the same for both loading paths; they are governed by the gradient of the failure envelope at A and not by the loading path.

The behaviour of an ideal elastic–perfectly plastic material can be represented by the behaviour of the simple model illustrated in Fig. 3.11(a). This consists of a soft rubber block with a frictional sandpaper base and a rigid platen bonded to the top. A constant normal force F_n and variable horizontal forces F_x and F_y are applied to the platen. If the horizontal forces are less than required to cause frictional sliding of the sandpaper over the table all deformations of the platen are due to elastic deformation of the rubber block. Thus increments of force $\pm\delta F_x$ cause displacements $\pm\delta x^e$ in the direction of the force as shown in Fig. 3.11(b). If, however, there is frictional sliding then the direction of plastic (irrecoverable) displacement δ^p is in the direction of the resultant force **F** and is independent of the increment of load δF_x or δF_y, as shown in Fig. 3.11(c).

3.7 COMBINED ELASTO–PLASTIC BEHAVIOUR

With reference to Fig. 3.1, the stress–strain behaviour is elastic up to the yield point and is perfectly plastic at the ultimate state. Between the first yield and failure there are simultaneous elastic and plastic components of strain.

(a) Simple model

(b) Elastic deformations

(c) Plastic deformations

Figure 3.11 A physical model for elastic and plastic behaviour.

Figure 3.12 Material behaviour during load cycling.

In Fig. 3.12 material is loaded from O_1 and is elastic until yielding occurs at Y_1, where the yield stress is σ'_{x1}. It is then strained further and unloaded to O_2 where there are irrecoverable plastic strains $\delta\varepsilon^p_{x1}$. When the material is reloaded from O_2 it is elastic until yielding occurs at Y_2, where the yield stress is σ'_{x2}. If the material is then strained further and unloaded to O_3, on reloading it will have a new yield stress σ'_{x3} and so on. Thus the principal consequences of straining from Y_1 to Y_2 (or from Y_2 to Y_3) are to cause irrecoverable plastic strains and to raise the yield point from σ'_{x1} to σ'_{x2} (or from σ'_{x2} to σ'_{x3}). This increase of the yield point due to plastic straining is called hardening and the relationship between the increase in the yield stress $\delta\sigma'_x$ and the plastic straining $\delta\varepsilon^p_x$ is known as a hardening law. In Fig. 3.12 there is a broken

(a) Strain hardening

(b) Strain softening

Figure 3.13 Yielding and plastic straining.

line to the left of the first yield point which suggests that there could be even lower yield points for previous loadings; this simply demonstrates that the origin of strains O_1 was arbitrary chosen.

Yielding and plastic straining may cause hardening (i.e. an increase in the yield stress), as shown in Fig. 3.13(a), or softening (i.e. a decrease in the yield stress), as shown in Fig. 3.13(b). In the latter case the state has reached, and passed, a peak in the stress–strain curve, and this is a feature commonly found in the behaviour of soils. In each case the total strains are the sum of the elastic and plastic components and the plastic strains are related to the change of the yield stress by a hardening law.

Yielding under combined stresses may be represented by a set of yield curves which are similar to the failure envelope, as illustrated in Fig. 3.14. This shows a yield curve for the first yield, two yield curves for subsequent yielding and a failure envelope. For states inside the first yield curve the behaviour is elastic. The state cannot reach the region outside the failure envelope. If the plastic strains are perfect then the vectors of plastic strain are normal to the yield curves. Thus, for the loading path A → B in Fig. 3.15 which crosses successive yield surfaces the vectors of plastic strain are normal to the yield surface.

Since each yield curve in Fig. 3.15 is associated with a particular plastic strain we can use the plastic strain as a third axis to develop a yield surface as shown in Fig. 3.16. For any state on the yield surface there are plastic strains that are normal to the appropriate yield curve and are given by the movement of the stress point across the surface. For any state inside the surface,

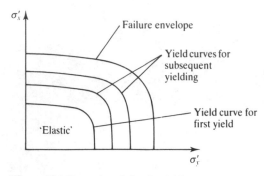

Figure 3.14 Examples of simple yield curves.

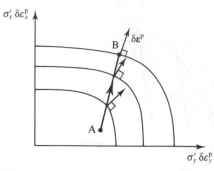

Figure 3.15 Plastic straining for loading on a yield surface.

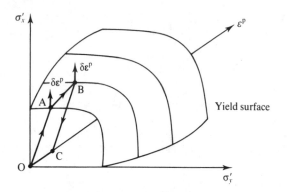

Figure 3.16 Behaviour during a cycle of loading on and under the yield surface.

during first loading or due to unloading, the behaviour is elastic. Thus, for the loading and unloading O → A → B → C in Fig. 3.16 the behaviour is elastic for the paths O → A and B → C. For the path A → B there are simultaneous elastic and plastic strains.

It is now possible to assemble the flow rule, the hardening law and the elastic stress–strain equations into an explicit constitutive equation for the complete range of loading up to failure. We will develop such a constitutive equation for soil in Chapter 12 when we have obtained equations for the yield surface and for the successive yield curves for soil.

3.8 TIME AND RATE EFFECTS

In developing constitutive equations for materials we have, so far, considered only relationships between changes of effective stress and changes of strain. This means that no strains occur at constant load (except at failure). In addition it was assumed that the relationships between stress and strains were independent of the rate of loading or the rate of straining. In soils there are a number of time and rate effects mainly due to drainage of water and, to a limited extent, due to creep and viscosity in the soil skeleton.

Time-dependent straining due to drainage of water is known as consolidation and it is a coupling of deformations due to effective stress with seepage. Theories for consolidation will be considered in Chapter 14.

The theory of viscosity relates stresses in moving materials (usually fluids) to the velocity of flow, so that the shear stresses in water flowing in a pipe are related to the velocity of the flow. In solid materials such as steel, concrete or soil, the strength or stiffness may be governed by the rate of loading or by the rate of straining. It turns out that the important mechanical properties of most soils, except peats and organic soils, are not significantly influenced by the rate of loading, and usually we will not have to worry about viscous effects in soil mechanics.

Materials under constant stress generally continue to strain, but at a rate that diminishes with time; this is known as creep. The basic relationship for creep is

$$\delta\varepsilon^c = C_\alpha \ln(t/t_0) \tag{3.26}$$

where C_α is a creep parameter that depends on a number of factors, including the magnitudes of the (constant) stresses, and t_0 is time from which the creep strains are measured. Equation (3.26) can be differentiated to give the creep strain rate as

$$\frac{d\varepsilon^c}{dt} = \frac{C_\alpha}{t} \tag{3.27}$$

showing that the creep strain rate decreases with time. Many soils, particularly soft clays and peats, show significant creep strains. These can also influence the subsequent behaviour, as we will discuss later.

3.9 SUMMARY

1. The basic mechanical properties of structural materials are stiffness and strength. Stiffness relates changes of stress to changes of strain and this governs deformations and ground movements. Strength is the largest shear stress that a material can sustain before it fails and this governs the ultimate states of collapse of structures.
2. Strength can be described either by the Tresca criterion $\tau_f = s$ or by the Mohr–Coulomb criterion $\tau_f' = c' + \sigma' \tan \phi'$.
3. Purely elastic strains are recovered on unloading. In metals the elastic stress–strain line is approximately linear so the elastic parameters G' and K' are approximately constants.
4. A perfectly plastic material continues to strain with constant stresses and the vector of plastic strain, which relates the rates of plastic shear and volumetric strains, is normal to the current yield curve.
5. Theories for elasto-plastic straining can be obtained by adding the elastic and plastic components of strain.
6. Most time and rate effects in soils are due to coupling of stiffness with seepage of pore water. Creep and viscous effects are usually neglected except in peats and other organic soils.

WORKED EXAMPLES

Example 3.1: Stress and strain in a triaxial test In a triaxial compression test on a sample of soil the pore pressure is zero so total and effective stresses are equal. The radial stress is held constant at $\sigma_r' = 200\,\text{kPa}$ and the axial stress is changed from $\sigma_a' = 350\,\text{kPa}$ to $360\,\text{kPa}$. The strains for this increment were $\delta\varepsilon_a = 0.05$ per cent and $\delta\varepsilon_r = -0.01$ per cent.

At the start of the increment,

$$q' = \sigma_a' - \sigma_r' = 350 - 200 = 150\,\text{kPa}$$
$$p' = \tfrac{1}{3}(\sigma_a' + 2\sigma_r') = \tfrac{1}{3}(350 + 400) = 250\,\text{kPa}$$

During the increment $\delta\sigma'_a = 10 \, \text{kPa}$, $\delta\sigma'_r = 0$ and, from Eqs (3.5) to (3.8),

$$\delta q' = (\delta'_a - \delta\sigma'_r) = 10 \, \text{kPa}$$

$$\delta p' = \tfrac{1}{3}(\delta\sigma'_a + 2\sigma'_r) = \tfrac{1}{3} \times 10 = 3.3 \, \text{kPa}$$

$$\delta\varepsilon_s = \tfrac{2}{3}(\delta\varepsilon_a - \delta\varepsilon_r) = \tfrac{2}{3}(0.05 + 0.01) = 0.04 \text{ per cent}$$

$$\delta\varepsilon_v = \delta\varepsilon_a + 2\delta\varepsilon_r = 0.05 - 0.02 = 0.03 \text{ per cent}$$

Example 3.2: Calculation of shear and bulk modulus The soil in Example 3.1 is isotropic and elastic (i.e. shearing and volumetric effects are decoupled). For the increment,

$$\text{shear modulus } G' = \frac{\delta q'}{3\delta\varepsilon_s} = \frac{10}{3 \times 0.04/100 \times 1000} = 8.3 \, \text{MPa}$$

$$\text{bulk modulus } K' = \frac{\delta p'}{\delta\varepsilon_v} = \frac{3.3}{0.03/100 \times 1000} = 11.1 \, \text{MPa}$$

$$\text{Young's modulus } E' = \frac{\delta\sigma'_a}{\delta\varepsilon_a} = \frac{10}{0.05/100 \times 1000} \, 20 \, \text{MPa}$$

$$\text{Poisson's ratio } v' = -\frac{\delta\varepsilon_r}{\delta\varepsilon_a} = \frac{0.01}{0.05} = 0.2$$

From Eqs (3.24) and (3.25), substituting for E' and v',

$$G' = \frac{E'}{2(1 + v')} = \frac{20}{2(1 + 0.2)} = 8.3 \, \text{MPa}$$

$$K' = \frac{E'}{3(1 - 2v')} = \frac{20}{3(1 - 0.4)} = 11.1 \, \text{MPa}$$

FURTHER READING

Atkinson, J. H. (1981) *Foundations and Slopes*, McGraw-Hill, London.

Calladine, C. R. (1969) *Engineering Plasticity*, Pergamon Press, London.

Case, J. and A. H. Chilver (1971) *Strength of Materials and Structures*, Edward Arnold, London.

Jaeger, J. C. (1969) *Elasticity, Fracture and Flow*, Methuen, London.

Naylor, D. J. (1978) 'Stress–strain laws for soil', in *Developments in Soil Mechanics*, C. R. Scott (ed.), Applied Science Publishers, London.

Palmer A. C. (1976) *Structural Mechanics*, Oxford University Press.

Timoshenko, S. P. and J. N. Goodier (1951) *Theory of Elasticity*, McGraw-Hill, New York.

CHAPTER FOUR

THE STRUCTURE OF THE EARTH

4.1 INTRODUCTION

Soils occur very near the surface of the Earth and are essentially the products of the action of the weather and the climate on rocks. Weathering of rock *in situ* leads to the formation of residual soils. These may be eroded, transported and laid down as deposited soils. The engineering properties of soils and how they occur in the ground depend to a great extent on their geological origins and so geotechnical engineers will need to know something about geology.

In this one chapter I cannot possibly cover the whole of geology, or even all the parts related to engineering. You will find a number of simple and easy-to-read books on geology for engineers and on engineering geology and you will probably attend lectures on the subject. What I want to do here is set down what I consider to be the most interesting and important aspects of geology related to geotechnical engineering in soils. This is my personal list and other geotechnical engineers and geologists will probably want you to know about other things. This does not really matter because if you want to be a good geotechnical engineer you will need to study geology in some detail.

4.2 THE EARTH'S CRUST

The Earth has a radius of about 8000 km and a crust of soils and rocks about 25 to 50 km thick (see Fig. 4.1a). The ratio of the thickness of the crust to its radius of curvature is about the same as that of an eggshell. Below the crust is a mantle of hot plastic material and plates of crust move about on the mantle. This drift of the continental crust accounts for mountain building, earthquakes and volcanic activity at boundaries between the plates. It also accounts for evidence of glacial deposits in Australia and tropical soils in Antarctica. In a single core of rock taken almost anywhere on Earth, there will be rocks deposited in conditions that were like all the known present-day environments.

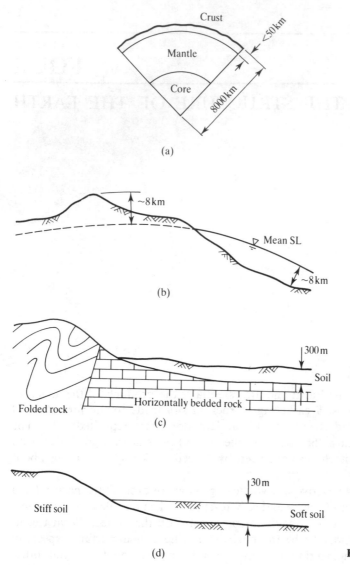

Figure 4.1 The structure of the Earth.

The surface of the crust (i.e. the land surface and the sea bed) has altitudes and depths above and below mean sea level of the order of 8 km (see Fig. 4.1b). Materials near the surface are soils and rocks although there is not a very clear distinction between the two; at low stresses soils fracture like rocks while at high stresses rocks will deform plastically like soils. For engineering purposes soils rarely occur below a depth of about 300 m (see Fig. 4.1c). Geologically old soils (older than about 2 millions years) are usually relatively stiff and strong while young soils (Glacial and Post-Glacial) are usually relatively soft and weak, but these are rarely deeper than about 30 m (see Fig. 4.1d). Notice that the slope of the land reflects the strength of the underlying material; in rocks mountain slopes can be steep and high while in soils the slope angles are much more gentle and the heights are much less. Spread out over most of the land surface is a layer of soil of variable thickness, but usually less than 1 m, that supports plant life. This is called topsoil; it is of great interest to farmers and gardeners but not to engineers, except to save and replace as landscaping after construction is complete.

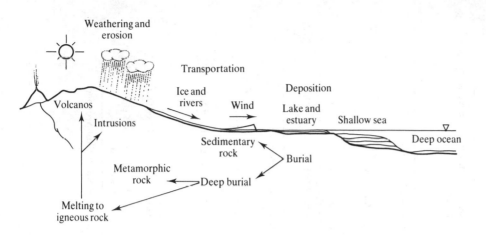

Figure 4.2 Simplified representation of the cycling of rocks and soils.

4.3 GEOLOGICAL PROCESSES

Soils and rocks close to the surface of the Earth are exposed to the atmosphere and are weathered, eroded, transported and deposited, while deep burial converts soils back to rocks. The general cycle of soils and rocks is illustrated in Fig. 4.2.

Collisions between drifting continental plates raise mountain chains like the Andes, the Rockies and the Himalayas. Rain, snow and sunshine weather rocks and soils; this may consist simply of mechanical breakdown of a rock mass into chunks of material that have the same composition as the parent rock or chemical alteration to new clay minerals. Water, ice and wind transport these weathered rock fragments and, at the same time, degrade, polish and sort them into different sizes. When these transporting agents slow down, the soil particles are deposited and as further material is deposited above they become compressed.

The cycle illustrated in Fig. 4.2 is, of course, highly simplified and there are many additional influences and processes. For example, rocks weathered *in situ* form residual soils while tectonic activity recycles molten material from below the crust to the surface, sometimes causing rocks and soils to metamorphose in the process.

4.4 STRATIGRAPHY AND THE AGE OF SOILS AND ROCKS

A borehole drilled down into the crust will pass through strata, or layers, of soils and rocks which generally become older with depth. Stratigraphy is the study of the sequence of strata that represent geological history. At a particular location there will have been periods of volcanic activity, mountain building and erosion and so a single borehole will not reveal the complete sequence of Earth's history. A break in the stratigraphic column in a borehole record is known as an unconformity; often the materials at an unconformity have been eroded before deposition of new material.

A highly simplified version of the stratigraphic column is shown in Table 4.1; this gives the name and approximate age of the major divisions, the general nature of the deposits and typical examples from the United Kingdom. In other parts of the World, the major divisions and their ages are the same but the nature of the deposits may well be different: for example, the Cretaceous Chalk in South East England is the same age as the Deccan Lavas in India.

In order to describe the chronological history of the Earth geologists classify major strata

Table 4.1 Simplified stratigraphical column in Britain

Name of geological group or era		Name of geological system or period (approx. age in millions of years)	General nature of deposits. Major orogenies (mountain building) and igneous activity (*in italics*)	Examples in the United Kingdom
CENOZOIC (= recent life)	Quaternary	Recent	Alluvium, blown sand, glacial deposits, etc.	Extensive drift cover to most of the bedrock
		Pleistocene	At least five major ice ages separated by warmer periods	
		(2)		
	Tertiary	Pliocene Miocene Oligocene Eocene	Sand, clays and shell beds *Alpine orogeny* *Igneous activity in west Scotland and Northern Ireland*	East Anglian Crags London Clay
		(65)		
MESOZOIC (= middle life)		Cretaceous Jurassic Triassic	Sand, clays and chalk Clays, limestones, some sands Desert marls, sandstone marls	Chalk, Gault Clay, Weald Clay Lias Clay, Portland Stone Keuper Marl, Bunter Sandstone
		(250)		
PALAEOZOIC (= ancient life)	Newer	Permian	Breccias, marls, dolomitic limestone *Hercynian orogeny* *Igneous activity*	Magnesian Limestone
		Carboniferous	Limestones, shales, coals and sandstones	Coal Measures and Mountain Limestone
		Devonian (and Old Red Sandstone)	Siltstones, sandstones and limestones (Lacustrine sandstones and marlstones) *Igneous activity* *Caledonian orogeny*	Red beds in Hereford and S. Wales
		(400)		
	Older	Silurian Ordovician Cambrian	Thick shallow-water limestones, shales and sandstones *Volcanic activity in the Ordovician* Slates, siltstones, sandstone and greywackes	Rocks of Wales and Lake District Rocks of North Wales, Central and North Scotland
		(600)		
PRE-CAMBRIAN		Torridonian Uriconian Lewisian	Schists Sandstones Lavas and tuffs (Shropshire) *Pre-Cambrian orogenies* Orthogneiss	Rocks of North West Scotland

according to their age, not what they are. Notice that the initial letters of geological names are capitals (e.g. Old Red Sandstone, London Clay, etc.) whereas the engineering descriptions (e.g. overconsolidated clay) have lower-case initial letters. For example, the deposit called London Clay is of Eocene age and was deposited 40 to 60 million years ago. The deposit is found in South East England and is also found in Belgium, where it is called Boom Clay. In the London region it is largely a marine clay but to the west of London, in the Hampshire Basin, it is mostly silt and fine sand with very little clay. Old Red Sandstone is of Devonian age and was deposited 350 to 400 million years ago. It is generally red in colour, unlike the Carboniferous rocks above and the Silurian rocks below, which are both grey, but it is not all sandstone and it contains great thicknesses of mudstones and siltstones.

Generally soils and rocks become stiffer and stronger with age: London Clay is obviously stronger than the soils found in the English Fens and the slates in North Wales are stronger still. As a very rough guide, materials of Cenozoic age are generally regarded as soils for engineering purposes; materials of Mesozoic age are generally regarded as soft rocks and materials of Palaeozoic age are regarded as hard rocks. The soils and rocks in the stratigraphic column contain fossils which are the most important indicators of their age and provide a record of evolution on Earth. Cambrian and Ordovician rocks contain mollusc shells and corals; land plants occur in the Devonian, reptiles in the Carboniferous, amphibians in the Permian, dinosaurs in the Triassic and birds in the Jurassic; the dinosaurs became extinct in the Cretaceous. Mammals, fishes, insects and birds had evolved by the Eocene, but modern man did not evolve until the middle of the Pleistocene, about 1 million years ago.

Since the engineering properties of sands, silts and clays and of sandstones, siltstones and mudstones are likely to be different, the standard geological age-based classifications will only be of limited use in geotechnical engineering. Much better schemes for engineering classifications of soils and rocks are based on the nature of the grains and on the state of stress and water content. These are described in Chapter 5.

4.5 DEPOSITIONAL ENVIRONMENTS

The nature of the weathering and the mode of transport largely determine the nature of a soil (i.e. the size and shape of the grains, the distribution of grain sizes and their mineralogy). The environment into which it is deposited and the subsequent geological events largely determine the state of the soil (i.e. the denseness or looseness of the packing of the grains) and its fabric (i.e. the presence of structural features such as layering, fissuring, bedding and so on).

As you move about the world you can see weathering, erosion, transportation and deposition taking place. In the present day in the United Kingdom most of the transportation is by water (rivers look dirty because they are carrying soil particles) and most of the deposition is in lakes, estuaries and in the near-shore region of the sea bed. In the past there have been many different climates and environments, because what is now the United Kingdom moved about the Earth on a drifting continental plate. Today in cold regions and at high altitudes you can seen transportation by glaciers and in deserts by wind, while in the tropical regions there are deep deposits of residual soils formed *in situ*. The study of depositional environments is a fascinating subject and is the key to the understanding and interpretation of engineering ground investigations. The basic principles are that all soils and rocks were deposited in one of a relatively small number of depositional environments, all of which can be found somewhere in the world today; if you know what the depositional environment was then you can infer much about the likely nature and structure of a deposit.

Figure 4.3 illustrates three typical depositional environments. Figure 4.3(a) shows the end of a moving glacier transporting eroded soil and rock. It deposits a basal till and a terminal

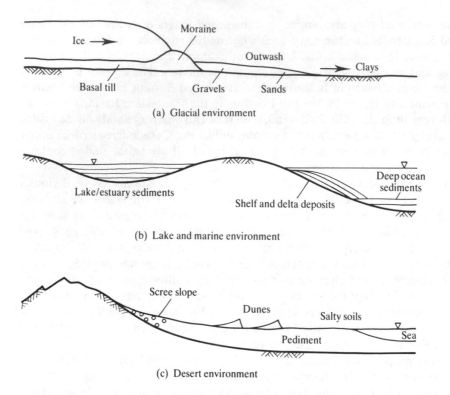

(a) Glacial environment

(b) Lake and marine environment

(c) Desert environment

Figure 4.3 Characteristic depostional environments.

moraine; the soils in both these deposits are well graded (i.e. they contain a wide variety of particle sizes from clays to boulders and they are often called boulder clay). Water from the melting glacier transports material away from the glacier but sorts the sizes, depositing first gravels, then sands and moving clays considerable distances. Figure 4.3(b) shows deposition into lakes or estuaries or into the oceans. Slow flowing rivers can only carry fine-grained soils, so the deposits will be largely silts and clays. Still water deposits tend to be layered horizontally while delta and moving water deposits are built in steps. Figure 4.3(c) shows a desert environment. Hot days and cold nights cause thermal weathering of rock mountains which produces scree slopes. Rare flash floods transport material across the desert floor (or pediment), depositing coarse material first and fine material later, probably in fans and layers. Winds cause migration of sand dunes. Large daily temperature variations with occasional rainfall cause physical and chemical alteration of the soils in the pediment.

These are only three typical depositional environments. They are discussed in more detail, together with other examples, by Fookes and Vaughan (1986). Much of the United Kingdom north of a line from the Thames to the Severn estuaries is covered with a veneer of glacial deposits. Most natural and man-made lakes are currently collecting layered silt and clay deposits. Large rivers (e.g. Nile, Ganges, Mississippi) are currently building delta deposits. Modern deserts occur widely throughout Asia, Australia, Africa, North and South America and the Middle East. Glacial environments occur in high latitudes (e.g. Greenland, Antarctica) and at high altitudes.

These typical depositional environments can be recognized in ancient rocks. For example, the London Clay was deposited in a shallow sea; the Chalk is calcium carbonate deposited in a warm sea; the New Red Sandstone in the Triassic and the Old Red Sandstone in the Devonian are ancient desert deposits. The important point to make here is that you should study present-day depositional environments as an aid to interpretation of ground investigations; if a geologist can tell you the environment into which a soil or rock was deposited you have a very good idea of what to expect.

4.6 RECENT GEOLOGICAL EVENTS

Although the depositional environment has a major influence on the formation of soils and rocks, they are altered by later geological events such as further deposition or erosion, folding and faulting and volcanic activity. For soils and soil mechanics the most significant recent geological events are rising or falling land and sea levels which lead to continuing deposition or erosion.

Land and sea levels rise and fall relative to one another for a variety of reasons, including plate movements and mountain building. One of the most important causes of changes of sea level is temperature change. During an ice age the sea cools and contracts and ice remains on the land as glaciers; the weight of ice depresses the land which rebounds as the ice melts. At the end of the last ice age, about 20 000 years ago, the sea level was about 100 m lower than it is now, so the UK coastline was west of Ireland and you could have walked to France (if you could cross the large river flowing through the Straits of Dover).

Figure 4.4 Stages of erosion and deposition during changing seas levels.

During the period of rising sea levels (e.g. at the end of an ice age) soils are deposited around the coasts. If the sea level remains stationary for some time vegetation grows, which is submerged and decays to peat as the sea level rises again. In the United Kingdom extensive deposits of this kind are found in the Wash and in the Somerset Levels. Continuing sea level rise and deposition leads to deposits of normally consolidated soils (see Chapter 15) which are soft and weak near the surface but become stronger with depth. During a period of falling sea level (e.g. at the beginning of an ice age) the land becomes exposed and subject to weathering, erosion and transportation. As the ground is eroded the soils become overconsolidated (see Chapter 15) due to unloading, but they do not recover their original state. Overconsolidated soils have stiffnesses and strengths which are more or less uniform with depth and which are larger than those of normally consolidated deposits at the same depth.

Figure 4.4 illustrates a sequence of falling and rising sea levels; this is a highly simplified model of the recent geology of the Thames estuary north east of London. In Fig. 4.4(a) London Clay is deposited in a shallow sea; notice an element at A just above the Chalk. Figure 4.4(b) shows a glaciation with a very low sea level and a nearby glacier. Much of the London Clay has by now been eroded so the element A is nearer the surface. Meltwater from the glacier has eroded a river channel which has been partly filled with outwash gravels. Figure 4.4(c) shows the present day; rising sea levels have led to deposition of soft soils in the valley. The soil at B is about the same depth as that at A, but it is normally consolidated and so is relatively soft and weak.

4.7 IMPORTANCE OF GEOLOGY IN GEOTECHNICAL ENGINEERING

It is obvious that an understanding of the geology of a location will aid the interpretation of ground investigations. All soils were deposited or formed *in situ* in one of only a few characteristic environments. These environments, together with later geological events, determine the nature and state of soils and rocks. Very nearly all the environments that have occurred on Earth can be found somewhere in the world today. What you have to do is find a geologist who can identify the geological environment of a deposit for you. Do not ask the geologist what the soil or rock is as you can usually see what it is yourself; instead, ask how did it get there (i.e. how was it deposited) and what has happened to it since.

FURTHER READING

Blyth, F. G. H. and M. H. De Freitas (1974) *A Geology for Engineers*, Edward Arnold, London.
Flint, R. F. (1971) *Glacial and Quaternary Geology*, Wiley, New York.
Fookes, P. G. and P. R. Vaughan (1986) *A Handbook of Engineering Geomorphology*, Surrey University Press.
Gass, I. G., P. J. Smith and R. C. L. Wilson (1971) *Understanding the Earth*, Open University Press, Sussex.
Holmes, A. H (1965) *Principles of Physical Geology*, Nelson, London.
ICE (1976) *Manual of Applied Geology for Engineers*, ICE, London.
McLean, A. C. and C. D. Gribble (1985) *Geology for Civil Engineers*, Chapman & Hall, London.
Read, H. H. and J. Watson (1966) *Beginning Geology*, Macmillan, London.
West, R. G. (1968) *Pleistocene Geology and Biology*, Longman, London.

CLASSIFICATION OF SOILS

5.1 DESCRIPTION AND CLASSIFICATION

Soils consists of grains, usually rock fragments or clay particles, with water and gas, usually air or water vapour, in the void spaces between the grains. If there is no gas present the soil is saturated and if there is no water it is dry, while if there is both water and gas in the voids the soil is unsaturated. The mechanics of unsaturated soils is very complicated and in this book I will consider only saturated or dry soils. Fortunately, in civil engineering applications soils are mostly saturated, except in hot dry environments or when compacted.

The mechanical properties of a soil (i.e. its strength and stiffness) depend principally on the nature of the grains (i.e. what they are) and the state of the soil (i.e. how the grains are arranged). You can dig up a sample of soil from your garden or from the beach and describe what you see. You can describe its colour, the size and shape of the grains (if you can see them) and some aspects of the behaviour, such as its response to moulding in your fingers. To be useful, however, you will need a scheme of classification that separates groups of soils with markedly different behaviour. Any useful scheme of soil classification should be based on relatively simple tests and observations.

It is important to distinguish between soil description and soil classification. Description is simply what you see and how the soil responds to simple tests; you may want to describe only a single soil sample or a soil profile exposed in a cliff face, in an excavation or from a number of samples from a borehole. A classification is a scheme for separating soils into broad groups, each with broadly similar behaviour. There are various classification schemes for different purposes: there are agricultural classifications based on how soils support crops and geological classifications based on the age of the deposit or the nature of the grains. For civil engineering purposes soil classifications should be based mainly on mechanical behaviour.

5.2 DESCRIPTION OF SOILS

Soil description is essentially a catalogue of what the soil is and it is helpful to have a simple scheme to describe the essential features. There are several such schemes published in National

Standards and to some extent these reflect the characteristics of the most common soils in the region; you should look up the relevant standard for the region you will work in. In the United Kingdom these are the British Standards for site investigations (BS 5930:1981) and for soil testing (BS 1377:1991) but slightly different schemes are used in the United States (Wagner, 1957). A simple and universal scheme for soil description is as follows:

1. The nature of the grains. The most important features of soil grains are their size and the grading (i.e. the proportions of different sizes), together with the shape and surface texture of the grains and their mineralogy.
2. The current state of the soil. The important indicators of the state of a soil are the current stresses, the current water content and the history of loading and unloading: these are reflected by the relative strengths and stiffnesses of samples of the soil.
3. The structure or fabric. Natural soils are rarely uniform. They contain features that may be on a scale of a few millimetres and observable in small samples or on a large scale and observable only in relative large exposures. Fabric or structure includes layering or bedding, fissuring or jointing and cementing.
4. The formation of the soil. Soils are formed in different ways. They may be deposited naturally from water, ice or wind; they may be the residual products of rock weathering; they may be compacted by machines into embankments and fills.

A more detailed scheme for description of soils is given in Table 5.1 which is taken from BS 5930:1981. This is similar to the scheme described above but is more detailed and gives helpful quantitative values for a number of visual observations. Notice the descriptions of compactness and strength and of structure, including guidance for descriptions of the spacing of bedding and discontinuities.

The nature of a soil does not usually change during normal civil engineering works; occasionally weak and brittle soil grains may fracture during loading so the grading changes. On the other hand, the state of a soil does change as soils near foundations and excavations are loaded or unloaded and compress or swell.

The manner of formation of a soil will influence both its nature, its initial state and its structure and fabric. Structure and fabric (i.e. layering, fissuring and jointing) can have an important influence on soil stiffness and drainage. In this book I will be examining the basic behaviour of soils observed in remoulded and reconstituted samples where any structure and fabric has been removed by the preparation of the sample. Since most natural soils have some structure and fabric it is important always to test some intact samples, but their behaviour should be examined within the basic framework established for reconstituted, destructured samples.

5.3 SOIL PARTICLE SIZES, SHAPES AND GRADINGS

The range of particle sizes in soils is very large and ranges from clay grains that are smaller than 2 μm (0.002 mm) to boulders that are larger than 200 mm. A particular range of particle sizes is given a name, as in Fig. 5.1 and Table 5.1, so that, for example, in UK practice medium sand is 0.2 to 0.6 mm. As a general guide, individual sand-sized and coarser particles are visible to the naked eye while individual silt-sized particles are visible using a × 10 hand lens. If you can wash fine grained soil off your boots it is probably silt, but if you have to scrape it off it is probably clay; similarly, if silt dries on your hands it will dust off while dry clay will leave your hands dirty and will have to be washed off. Further guidance for identification of sizes is given in Table 5.1.

Figure 5.1 Soil particle sizes

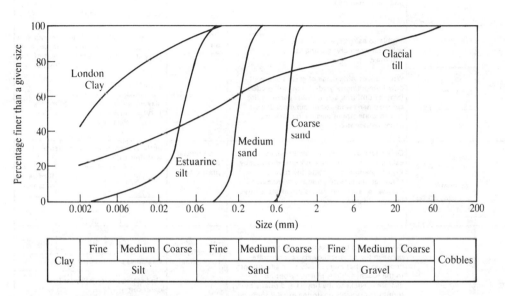

Figure 5.2 Grading curves plotted on a particle size distribution chart. (After BS 1377:1991.)

Soil particle shapes also differ considerably. Clay grains are usually very thin plates while silt, sand and gravel grains are more rotund.

Words such as sand, silt and clay are used both to classify a particular grain size and to describe a soil which may contain lesser quantities of other sizes. The distribution of particle sizes in a soil is represented by a grading curve on a particle size chart, as shown in Fig. 5.2. If the grading curve is flat the soil contains a wide variety of different particle sizes and is known to engineers as well graded; if the curve is steep and one size predominates the soil is poorly graded. The grading of a soil often reflects its origin. Soils deposited by rivers or wind tend to be poorly graded while boulder clays and tills deposited from ice tend to be well graded with a wide distribution of sizes. Tests to determine the grading of soils are described in Sec. 7.3.

5.4 PROPERTIES OF FINE-GRAINED SOILS

The behaviour of a coarse-grained soil (i.e. silt-sized and coarser), is very like that of an assembly of different sized marbles, but clays differ in two respect. Firstly, some clay grains themselves

Table 5.1 Description of soils. (From BS5930:1981)

	Basic soil type	Particle size, mm	Visual identification	Particle nature and plasticity	Composite soil types (mixtures of basic soil types)	
Very coarse soils	BOULDERS		Only seen complete in pits or exposures.	Particle shape: Angular Subangular Subrounded Rounded Flat Elongate	**Scale of secondary constituents with coarse soils**	
		— 200			**Term**	**% of clay or silt**
	COBBLES		Often difficult to recover from boreholes.			
		— 60				
Coarse soils (over 65 % sand and gravel sizes)	GRAVELS	coarse	Easily visible to naked eye; particle shape can be described; grading can be described.		slightly clayey ⎱ GRAVEL or slightly silty ⎰ SAND	under 5
		— 20	Well graded: wide range of grain sizes, well distributed. Poorly graded: not well graded. (May be uniform: size of most particles lies between narrow limits; or gap graded: an intermediate size of particle is markedly under-represented.)		— clayey ⎱ GRAVEL or — silty ⎰ SAND	5 to 15
		medium				
		— 6			very clayey ⎱ GRAVEL or very silty ⎰ SAND	15 to 35
		fine		Texture:		
		— 2				
	SANDS	coarse	Visible to naked eye; very little or no cohesion when dry; grading can be described.	Rough Smooth Polished	Sandy GRAVEL ⎱ Sand or gravel and important second constituent of the Gravelly SAND ⎰ coarse fraction	
		— 0.6				
		medium	Well graded: wide range of grain sizes, well distributed. Poorly graded: not well graded. (May be uniform: size of most particles lies between narrow limits; or gap graded: an intermediate size of particle is markedly under-represented.)		(See 41.3.2.2) For composite types described as: clayey: fines are plastic, cohesive: silty: fines non-plastic or of low plasticity	
		— 0.2				
		fine				
		— 0.06				
Fine soils (over 35 % silt and clay sizes)	SILTS	coarse	Only coarse silt barely visible to naked eye; exhibits little plasticity and marked dilatancy; slightly granular or silky to the touch. Disintegrates in water; lumps dry quickly; possess cohesion but can be powdered easily between fingers.	Non-plastic or low plasticity	**Scale of secondary, constituents with fine soils**	
		— 0.02			**Term**	**% of sand or gravel**
		medium				
		— 0.006			sandy ⎱ CLAY or gravelly ⎰ SILT	35 to 65
		fine				
		— 0.002				
	CLAYS		Dry lumps can be broken but not powdered between the fingers; they also disintegrate under water but more slowly than silt; smooth to the touch; exhibits plasticity but no dilatancy; sticks to the fingers and dries slowly; shrinks appreciable on drying usually showing cracks. Intermediate and high plasticity clays show these properties to a moderate and high degree, respectively.	Intermediate plasticity (Lean clay)	— CLAY:SILT	under 35
					Examples of composite types	
				High plasticity (Fat clay)	(Indicating preferred order for description) Loose, brown, subangular very sandy, fine to coarse GRAVEL with small pockets of soft grey clay	
Organic soils	ORGANIC CLAY, SILT or SAND	Varies	Contains substantial amounts of organic vegetable matter.		Medium dense, light brown, clayey, fine and medium SAND Stiff, orange brown, fissured sandy CLAY	
	PEATS	Varies	Predominantly plant remains usually dark brown or black in colour, often with distinctive smell; low bulk density.		Firm, brown, thinly laminated SILT and CLAY Plastic, brown, amorphous PEAT	

Table 5.1 (cont.)

Compactness/strength		Structure		Interval scales	Colour
Term	**Field test**	**Term**	**Field identification**		
Loose	By inspection of voids and particle packing.	Homo-geneous	Deposit consists essentially of one type.	**Scale of bedding spacing**	Red
Dense		Inter-stratified	Alternating layers of varying types or with bands or lenses of other materials.		Pink
			Interval scale for bedding spacing may be used.		Yellow / Brown
Loose	Can be excavated with a spade; 50 mm wooden peg can be easily driven.	Hetero-geneous	A mixture of types.		Olive / Green
Dense	Requires pick for excavation; 50 mm wooden peg hard to drive.	Weathered	Particles may be weakened and may show concentric layering.		Blue / White / Grey / Black etc.
Slightly cemented	Visual examination; pick removes soil in lumps which can be abraded.				Supplemented as necessary with: Light / Dark / Mottled etc.
Soft or loose	Easily moulded or crushed in the fingers.	Fissured	Break into polyhedral fragments along fissures. Interval scale for spacing of discontinuities may be used.		and
Firm or dense	Can be moulded or crushed by strong pressure in the fingers.				Pinkish
Very soft	Exudes between fingers when squeezed in hand.	Intact	No fissures.		Reddish / Yellowish / Brownish etc.
Soft	Moulded by light finger pressure.	Homo-geneous	Deposit consists essentially of one type.	**Scale of spacing of other discontinuities**	
Firm	Can be moulded by strong finger pressure.	Inter-stratified	Alternating layers of varying types. Interval scale for thickness of layers may be used.		
Stiff	Cannot be moulded by fingers. Can be indented by thumb.	Weathered	Usually has crumb or columnar structure.		
Very stiff	Can be indented by thumb nail.				
Firm	Fibres already compressed together.				
Spongy	Very compressible and open structure.	Fibrous	Plant remains recognizable and retain some strength.		
Plastic	Can be moulded in hand, and smears fingers.	Amor-phous	Recognizable plant remains absent.		

Scale of bedding spacing

Term	Mean spacing, mm
Very thickly bedded	over 2000
Thickly bedded	2000 to 600
Medium bedded	600 to 200
Thinly bedded	200 to 60
Very thinly bedded	60 to 20
Thickly laminated	20 to 6
Thinly laminated	under 6

Scale of spacing of other discontinuities

Term	Mean spacing, mm
Very widely spaced	over 2000
Widely spaced	2000 to 600
Medium spaced	600 to 200
Closely spaced	200 to 60
Very closely spaced	60 to 20
Extremely closely spaced	under 20

Table 5.2 Approximate values for the specific surface of some common soil grains

Soil grain	Specific surface (m^2/g)	Activity
Clay minerals		
Montmorillonite	Up to 840	>5
Illite	65–200	≈ 0.9
Kaolinite	10–20	≈ 0.4
Clean sand	2×10^{-4}	—

may show significant volume changes as the loading and water content changes; this accounts for clays tending to crack as they dry. Secondly, particle surface effects become significant.

The surface of a soil grain carries a small electrical charge which depends on the soil mineral and may be modified by an electrolyte in the pore water. These charges give rise to forces between soil grains in addition to their self-weight. The magnitudes of the interparticle forces are proportional to the surface areas of the grains, while self-weight forces are proportional to the volumes of the grains. As particle sizes decrease the surface forces diminish with the square of the effective diameter, whereas the self-weight forces diminish with the cube; consequently the effects of surface forces are relatively more important in fine-grained than in coarse-grained soils.

The relative importance of the surface and self-weight forces may be described by the specific surface. This is defined as the total surface area of all grains in unit mass. Table 5.2 lists typical values for the specific surface of the three common clay minerals and of clean sand; the differences in the values of specific surface for sand and clay are very large.

In coarse-grained soils such as silt, sand and gravel, particle surface forces are negligible compared to their self-weight forces, so that dry sand will run through an hour-glass and form a cone at the base. Dry fine-grained materials, such as kitchen flour, behave differently and if you squash a handful of flour in your hand it will form a coherent lump. This is because as the grains become densely packed and the number of contacts in unit volume increases, the slight surface forces give rise to a small cohesive strength; the lump is easily broken because the cohesive strength is very small. We will see later that true cohesive strength in soils is usually negligible unless they are cemented by other materials.

5.5 SPECIFIC VOLUME, WATER CONTENT AND UNIT WEIGHT

Many important mechanical properties of soil depend on the closeness of the packing of the grains, so that loose soils (i.e. where there is a high proportion of voids) will be weaker and more compressible than dense soils. The state of a soil can be described by the specific volume v given by

$$v = \frac{V}{V_s} \tag{5.1}$$

where V is the volume of a sample containing a volume V_s of soil grains. Sometimes the voids ratio e is used instead of specific volume, where

$$e = \frac{V_w}{V_s} \tag{5.2}$$

and V_w is the volume of the voids which, in saturated soil, are filled with water. Since $V = V_w + V_s$,

$$v = 1 + e \tag{5.3}$$

For coarse-grained soils, where surface forces are negligible, the grains pack together like spheres. The maximum specific volume of a loose assembly of uniform spheres is 1.92 and the minimum specific volume of a dense assembly is 1.35; common sands and gravels have specific volumes in the range $v = 1.3$ to 2.0. For fine-grained clay soils surface effects may be significant, especially at low stresses, and the maximum specific volume of a recently sedimented clay will depend on the clay mineral and any electrolyte in the pore water. Montmorillonite clays with large specific surfaces may exist with specific volumes in excess of 10, while kaolinite clays which have smaller specific surfaces have a maximum specific volume around 3. Under large loads the specific volumes of clay soils may be reduced to as little as $v = 1.2$ as the flat clay plates become nearly parallel.

Specific volume cannot be measured directly but it can be calculated from other easily measured parameters. The most convenient is water content w, defined as

$$w = \frac{W_w}{W_s} \tag{5.4}$$

and unit weight γ defined as

$$\gamma = \frac{W}{V} \tag{5.5}$$

where W_w is the dry weight of water evaporated by heating soil to 105°C, W_s is the weight of dry soil, $W = W_w + W_s$ is the weight of a sample with volume V. Standard tests to measure water content and unit weight are described in Sec. 7.3. For a typical clay soil the water content might be in the range 0.20 to 0.70 (i.e. 20 to 70 per cent) and the unit weight might be 18 to 22 kN/m³ (i.e. about twice that of water: $\gamma_w = $ approximately 10 kN/m³).

Relationships between these and specific volume can be obtained from Fig. 5.3 together with Eqs (5.1) to (5.5) as

$$e = v - 1 = wG_s \tag{5.6}$$

$$\gamma = \left(\frac{G_s + v - 1}{v}\right)\gamma_w \tag{5.7}$$

where G_s is the specific gravity of the soil grains which, for many soils, is approximately $G_s = 2.65$.

	Volumes	Weights
Water	V_w	$W_w = \gamma_w V_w$
Grains	V_s	$W_s = \gamma_w G_s V_s$
Totals	V	W

Figure 5.3 Grains and water in saturated soils.

5.6 LIMITS OF CONSISTENCY

As the water content and specific volume of a soil are increased it will soften and weaken; this is well known to farmers and football players. If the water content is very large we just get muddy water and if it is very small we get a material that is very hard and brittle like rock. Obviously there are limits to the water content within which a soil has the consistency of soil rather than the consistency of a liquid or a brittle rock. Tests to determine the precise water contents at which soil behaviour becomes liquid or brittle are the Atterberg limits tests described in Sec. 7.3; these determine the liquid limit (LL) where the soil starts to flow like a liquid and the plastic limit (PL) where it ceases to be plastic and becomes brittle.

The Atterberg limits apply to fine-grained soils. (Soils for which it is possible to determine the Atterberg limits are often called plastic, but this term must not be confused with the strict meaning of plastic as a type of constitutive relationship, discussed in Sec. 3.5.) For coarse-grained sands and gravels the appropriate limits are the minimum density of a very loosely poured sample and the maximum density of a vibrated and heavily loaded sample (Kolbuszewski, 1948). Thus the minimum density of a sand is equivalent to the liquid limit of a clay, while the maximum density is equivalent to the plastic limit. The relationships between the Atterberg limits and the maximum and minimum densities are illustrated in Fig. 5.4.

An important parameter for clay soils is the plasticity index (PI), defined as

$$PI = LL - PL \tag{5.8}$$

This defines the range of water content for a soil and is related to the maximum volume change (or compressibility) of the soil. Similarly, the difference between the maximum and minimum densities is related to the relative compressibility. These limits depend on the grading and on the mineralogy, shape and surface texture of the grains, so they describe the nature of the soil. The Atterberg limits are measured on soil passing a 425 μm sieve and this fraction contains both

(a) Fine-grained soils

(b) Coarse-grained soils

Figure 5.4 Limits of consistency of soils.

clay and silt particles. The activity A is defined as

$$A = \frac{PI}{\% \text{ by weight of clay}} \tag{5.9}$$

This is closely related to the specific surface and to the mineralogy of the clay. Typical values for the common clay minerals are given in Table 5.2.

5.7 CURRENT STATE

Because soil is both frictional and relatively highly compressible its stiffness, strength and specific volume all depend on the current stresses and history of loading and unloading during deposition and erosion. In Fig. 5.5(a) the soil at a shallow depth z is lightly loaded by the small vertical stress σ_z due to the weight of soil above and it is loose. After deposition of a substantial depth of soil z_1 as in Fig. 5.5(b), the same soil is heavily loaded and has become dense. After erosion back to the original ground level, as in Fig. 5.5(c), the same soil is again lightly loaded but remains relatively dense. Thus the current water content or density of a soil will depend on the current stress and on the history of loading and unloading.

The current state of a soil can be related to the relative position with respect to the limiting states. For fine-grained clay soils the liquidity index (LI) is defined as

$$LI = \frac{w - PL}{LL - PL} \tag{5.10}$$

where w is the current water content and for coarse-grained soils the relative density (D_r) is defined as

$$D_r = \frac{v_{max} - v}{v_{max} - v_{min}} \tag{5.11}$$

where v is the current specific volume. These relationships are illustrated in Fig. 5.4. Notice that a liquidity index of 1.0 (corresponding to the loosest or wettest state) corresponds to a relative density of zero.

Figure 5.5 Changes of state during deposition and erosion.

Table 5.3 Strength of clay soils estimated from observations in hand samples

Consistency	Identification	Undrained strength s_u kPa
Very soft	Extrudes between fingers	<20
Soft	Easily moulded in fingers	20–40
Firm	Moulded by strong finger pressure	40–75
Stiff	Cannot be moulded in fingers	75–100
Very stiff	Brittle and very tough	>150

Another measure of the consistency of a clay soil is its immediate strength. We will see later that the (undrained) strength of a clay is related to the liquidity index, as illustrated in Fig. 5.4(a). When the water content of a clay soil is at its liquid limit the strength is close to 1.5 kPa and when the water content is at the plastic limit the strength is close to 150 kPa. Rapid estimates of the strength of clays can be made in hand samples using the criteria in Table 5.3.

5.8 ORIGINS OF SOILS

The mechanical behaviour of a soil is determined principally by its nature and its current state, but these are governed, to some extent, by the manner of formation of the soil which may be deposited, residual or compacted by machines. Detailed discussions of the influence of the manner of formation of soils on their nature and state are beyond the scope of this book and are contained in books on engineering geology, but there are a few simple observations to be made:

1. Deposited soils. Soils may be deposited from water, ice or wind and the grading and particle shape and texture are governed largely by the transporting agent. Soils deposited from water or air are poorly graded because the ability of rivers or wind to move different sizes depends on the velocity, while soils deposited from ice (i.e. boulder clays) are well graded because ice can move all particle sizes equally. Abrasion in moving water or air produces rounded and polished grains while soil grains transported by ice generally retain their original shape and texture. The mineralogy of transported soils is simply that of the parent material, which may be rock fragments or weathered and eroded clay. The fabric and structure of deposited soil is usually bedding and layering, reflecting changes in the depositional environment.
2. Residual soils. These are the products of weathering of rocks, or soils, *in situ*. Their grading and mineralogy depend in part on the parent material but principally on the depth and type of weathering and on details of the drainage conditions. Residual soils usually have low water contents and liquidity indices (or high relative density) and may be unsaturated. The fabric of immature residual soils often reflects the fabric of the parent rock.
3. Compacted soils. Soils may be compacted into fills by rolling, vibration or impact. They are usually unsaturated initially but may later become saturated. Often soils are compacted in layers and may show horizontal structure.

5.9 SIMPLE PRACTICAL EXERCISES

A description of soils in laboratory samples and *in situ* is a very important part of ground investigations and you should try this for yourself. As part of an undergraduate course you will probably carry out grading analyses and Atterberg limit tests. To aid visual assessment of grain size it is helpful to prepare a set of jars, each containing a particular grain size from fine sand to coarse gravel. You should also try a rapid grain size analysis using sedimentation in a bottle, as described in Sec. 7.3.

An important description of the state of a fine-grained soil is the consistency given in Tables 5.1 and 5.3. You should handle samples of the same soil with different water contents and different consistencies.

As a practical laboratory exercise you should describe a bulk sample of a coarse-grained or a well-graded soil and an intact sample of a fine-grained soil; for the latter the sample should be split to expose the structure and fabric of the soil. As a field exercise you should find a section in a quarry, in a cliff or in an excavation and prepare a detailed log for an imaginary borehole (see Sec. 16.7); before you start work be sure that the face is stable.

5.10 SUMMARY

Classification of soils requires a careful and detailed description of the soil *in situ* and in samples together with some simple classification tests. The important characteristics required for description of soils are:

1. The nature of the grains including the grading (i.e. the distribution of particle sizes) and the mineralogy, particularly of clay soils. The Atterberg limits give indications of clay mineralogy.
2. The state of the soil given by the stresses, the history of deposition and erosion and the water content. Important indicators of soil state are the liquidity index of fine-grained soils or the relative density of coarse-grained soils.
3. Structure and fabric features including bedding, layering, fissuring and jointing.
4. The method of formation of the soil, which may be deposited from water, wind or ice, residual formed by weathering or compacted by rolling, vibration or impact.

WORKED EXAMPLES

Example 5.1: Grading of soils Table 5.4 gives the results of particle size tests on three different soils. The grading curves are shown in Fig. 5.6.

Table 5.4

BS sieve	Size from sedimentation (mm)	% smaller		
		Soil A	Soil B	Soil C
63 mm			100	
20 mm			75	
6.3 mm		100	66	
2 mm		96	60	
600 μm		86	55	
212 μm		10	45	100
63 μm		2	34	95
	0.020		22	84
	0.006		15	68
	0.002		8	42

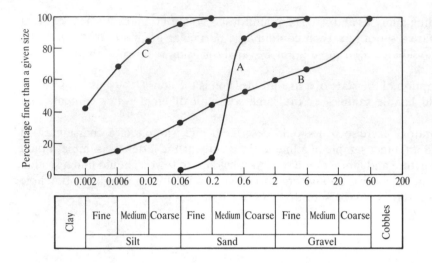

Figure 5.6

Soil A is predominantly sand; it is poorly (i.e. uniformly) graded with a relatively small range of sizes. It was probably deposited from a relatively fast flowing river. The permeability will be relatively large. The stiffness and strength will depend on the relative density and the current effective stresses.

Soil B is well graded with a very wide range of particle sizes from coarse gravel to fine silt with a little clay. It was probably deposited from a glacier and has not been sorted by wind or water. The permeability will be relatively low. *In situ* it is likely to have a low water content and, as a result, it will probably be relatively stiff and strong.

Soil C is a silty clay. It could be deposited either in a shallow sea, in a lake or in an estuary. Its stiffness and strength will depend on the mineralogy of the clay fraction as well as on the current water content and effective stress.

Example 5.2: Calculations of the state of a soil sample   A sample of saturated soil is 38 mm in diameter and 76 mm long and its mass is 142 g. After oven drying at 105°C its mass is 86 g.

$$\text{Water content } w = \frac{W - W_d}{W_d} = \frac{142 - 86}{86} = 0.651 = 65.1 \text{ per cent}$$

$$\text{Weight of saturated soil } W = 142 \times 9.81 \times 10^{-6} \text{ kN}$$

$$\text{Volume of cylinder } V = \frac{\pi}{4} \times 38^2 \times 76 \times 10^{-9} \text{ m}^3$$

$$\text{Unit weight } \gamma = \frac{W}{V} = 15.75 \text{ kN/m}^3$$

From Eqs (5.6) and (5.7)

$$\frac{\gamma}{\gamma_w} = \left(1 - \frac{1}{v}\right)\left(\frac{1}{w} + 1\right)$$

$$\frac{15.75}{9.81} = \left(1 - \frac{1}{v}\right)\left(\frac{1}{0.651} + 1\right)$$

$$v = 2.72$$

From Eq. (5.6),

$$G_s = \frac{v-1}{w} = \frac{1.72}{0.651} = 2.65$$

Example 5.3: Atterberg limits and soil mineralogy The Atterberg limits of a soil are LL = 70 and PL = 35 and it contains 80 per cent by weight of clay. The water content of a sample is 45 per cent.

$$\text{Plasticity index PI} = \text{LL} - \text{PL} = 70 - 35 = 35$$

$$\text{Liquidity index LI} = \frac{w - \text{PL}}{\text{PI}} = \frac{45 - 35}{35} = 0.29$$

$$\text{Activity } A = \frac{\text{PI}}{\% \text{ clay}} = \frac{35}{80} = 0.44$$

The clay is likely to be predominantly kaolinite.

Example 5.4: Calculation of the state of a soil A 1.5 kg sample of dry sand is poured into a Eureka can (see Fig. 5.7) and displaces 560 cm³ of water. The volume of the soil grains is equal

1.5 kg

560 cm³

(a)   (b)

Figure 5.7 Eureka can experiment.

to the volume of the water displaced from the can and so

$$\text{specific gravity } G_s = \frac{\text{weight of soil grains}}{\text{volume of soil grains}} = \frac{1.5 \times 10^3}{560} = 2.68$$

A second 1.5 kg sample of the same dry sand is poured into an empty measuring cylinder 55 mm in diameter and occupies 950 cm³ (see Fig. 5.8(a)). Therefore,

$$\text{specific volume } v = \frac{\text{volume of soil}}{\text{volume of grains}} = \frac{950}{560} = 1.70$$

$$\text{unit weight of dry soil } \gamma_d = \frac{\text{weight of dry soil}}{\text{volume}} = \frac{1.5 \times 9.81 \times 10^{-3}}{950 \times 10^{-6}} = 15.5 \text{ kN/m}^3$$

$$\text{depth of dry sand} = \frac{\text{volume}}{\text{area}} = \frac{950 \times 10^{-6}}{\pi/4 \times 55^2 \times 10^{-6}} = 0.40 \text{ m}$$

(a) (b) (c) **Figure 5.8**

When the cylinder is carefully filled with water up to the top level of the sand (see Fig. 5.8b),

$$\text{unit weight } \gamma = \left(\frac{G_s + v - 1}{v}\right)\gamma_w = \left(\frac{2.68 + 1.70 - 1}{1.70}\right)9.81 = 19.5 \text{ kN/m}^3$$

The side of the measuring cylinder is tapped several times, causing the level of the sand to settle to a volume of 870 cm³. At the new denser state (see Fig. 5.8c),

$$\text{specific volume } v = \frac{\text{volume of soil}}{\text{volume of grains}} = \frac{870}{560} = 1.55$$

$$\text{unit weight } \gamma = \left(\frac{G_s + v - 1}{v}\right)\gamma_w = \left(\frac{2.68 + 1.55 - 1}{1.55}\right)9.81 = 20.4 \text{ kN/m}^3$$

$$\text{depth of soil } z = \frac{\text{volume}}{\text{area}} = \frac{870 \times 10^{-6}}{\pi/4 \times 55^2 \times 10^{-6}} = 0.37 \text{ m}$$

REFERENCES

BS 5930 (1981) *Code of Practice for Site Investigations*, British Standards Institution, London.
BS 1377 (1991) *Methods of Test for Soils for Civil Engineering Purposes*, British Standards Institution, London.
Kolbuszewski, J. J. (1948) 'An experimental study of the maximum and minimum porosities of sands', *Proceedings of 2nd International SMFE Conference, Rotterdam*, Vol. 1.
Wagner, A. A. (1957) 'The use of the unified soils classification system by the Bureau of Reclamation', *Proceedings of 4th International SMFE Conference, London*, Vol. 1.

FURTHER READING

Atkinson, J. H. and P. L. Bransby (1978) *The Mechanics of Soils*, McGraw-Hill, London.
Clayton, C. R. I., N. E. Simons and M. C. Matthews (1982) *Site Investigation*, Granada, London.
Grimm, R. E. (1962) *Applied Clay Mineralogy*, McGraw-Hill, New York.
Head, K. H. (1980) *Manual of Soil Laboratory Testing*, Vol. 1, *Soil Classification and Compaction Tests*, Pentech Press, London.
Mitchell, J. K. (1976) *Fundamentals of Soil Behaviour*, Wiley, New York.

CHAPTER SIX

PORE PRESSURE, EFFECTIVE STRESS AND DRAINAGE

6.1 INTRODUCTION

Soils consist of solid grains and water, and loads on foundations or on walls will arise from combinations of the stresses in the skeleton of soil grains and in the pore water. If there is no soil the normal stress on the hull of a ship is equal to the water pressure. If there is no water the stress on the bottom of a sugar basin arises from the weight of the dry sugar. The question then arises as to what combinations of the stresses in the skeleton of the grains and in the pore water determine the overall soil behaviour. To examine this we will look at the stresses and water pressures in the ground.

6.2 STRESS IN THE GROUND

In the ground the vertical stress at a particular depth is due to the weight of everything above—soil grains, water, foundations—and so stresses generally increase with depth. In Fig. 6.1(a) the vertical stress σ_z is

$$\sigma_z = \gamma z \qquad (6.1)$$

Figure 6.1 Total stresses in the ground.

where γ is the unit weight of the soil (see Sec. 5.5). If the ground is below water level, in the bed of a lake or a sea, as in Fig. 6.1(b),

$$\sigma_z = \gamma z + \gamma_w z_w \tag{6.2}$$

and if there is a surcharge load q at the surface from a foundation or an embankment, as in Fig. 6.1(c),

$$\sigma_z = \gamma z + q \tag{6.3}$$

Remember that γ is the weight of everything (soil grains and water) in unit volume. Because σ_z arises from the total weight of the soil it is known as a total stress. Notice that the water in the lake in Fig. 6.1(b) applies a total stress at the ground surface in the same way that water in a glass applies total stresses to the bottom of the glass. The specific gravity of soil grains does not vary very much and, typically, $\gamma \approx 20 \text{ kN/m}^3$ for saturated soil, $\gamma \approx 16 \text{ kN/m}^3$ for dry soil and for water $\gamma_w \approx 10 \text{ kN/m}^3$.

There are also total horizontal stresses σ_h, but there are no simple relationships between σ_z and σ_h. We will examine horizontal stresses in later chapters.

6.3 GROUNDWATER AND PORE PRESSURE

The water in the pores of saturated soil has a pressure known as the pore pressure u. This is conveniently represented by the height of water h_w in a standpipe, as shown in Fig. 6.2. When everything is in equilibrium the pressures of water just inside and just outside the pipe are equal and so

$$u = \gamma_w h_w \tag{6.4}$$

When the level of water in the pipe is below ground, as in Fig. 6.2(a), it is known as the water table or the phreatic surface. If the water in the soil is stationary the water table is horizontal like the surface of a lake. However, as we will see later, if the phreatic surface is not level there will be seepage as the groundwater moves through the pores of the soil. From Fig. 6.2(a) pore pressures at the water table are zero (this is a definition of the phreatic surface) and positive below and a question is: what is the pore pressure above the phreatic surface?

Figure 6.3 illustrates the variation of pore pressure in the region between the ground level and the water table. There may be a layer of dry soil at the surface where pore pressures are zero. This is actually relatively rare but can be found on beaches above the high-tide mark. Immediately above the water table the soil remains saturated because of capillary rise in the

(a) (b) **Figure 6.2** Pore water pressures in the ground.

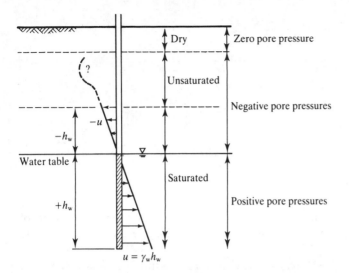

Figure 6.3 Pore pressures and suctions in the ground.

pore spaces. In this zone the pore pressures are negative and are given by

$$u = -\gamma_w h_w \tag{6.5}$$

An important point to notice is that saturated soils may very well have negative pore pressures. This implies that the water is in tension and the rise of water in soil above the phreatic surface is like the rise of water in a capillary tube. The height of the saturated region above the water table depends essentially on the size of the grains, or more particularly on the size of the pore spaces: the smaller the grains and pores the greater the height of saturated soil with negative pore pressures and the greater the magnitude of the greatest negative pore pressure at the top of the saturated zones.

Between the dry and saturated zones is a zone of unsaturated soil which contains soil grains, water and gas, usually air or water vapour. In this soil the pore water and the gas exist at different pressures and the pore water suctions may increase or decrease as indicated in Fig. 6.3. At present there is no simple and satisfactory theory for unsaturated soil and in this book I will only deal with dry or saturated soils. For practical purposes soils controlling the behaviour of slopes, foundations, retaining walls and other major civil engineering structures are usually saturated, at least in temperate or wet climates. Unsaturated soils occur in soils very near the surface, in compacted soils and in hot dry climates.

6.4 EFFECTIVE STRESS

It is obvious that ground movements and instabilities can be caused by changes of total stress due to loading of foundations or excavation of slopes. What is perhaps not so obvious is that ground movements and instabilities can be caused by changes of pore pressure. For example, stable slopes can fail after rainstorms because the pore pressures rise due to infiltration of rainwater into the slope while lowering of groundwater due to water extraction causes ground settlements. (Some people will tell you that landslides occur after rainfall because water lubricates soil; if they do, ask them to explain why damp sand in a sand-castle is stronger than dry sand.)

If soil compression and strength can be changed by changes of total stress or by changes

Figure 6.4 Mohr circles of total and effective stress.

of pore pressure there is a possibility that soil behaviour is governed by some combination of σ and u. This combination should be called the *effective* stress because it is effective in determining soil behaviour.

The relationship between total stress, effective stress and pore pressure was first discovered by Terzaghi (1936). He defined the effective stress in this way:

> All measurable effects of a change of stress, such as compression, distortion and a change of shearing resistance, are due exclusively to changes of effective stress. The effective stress σ' is related to the total stress and pore pressure by $\sigma' = \sigma - u$.

Figure 6.4 shows Mohr circles of total stress and effective stress plotted on the same axes. Since $\sigma'_1 = \sigma_1 - u$ and $\sigma'_3 = \sigma_3 - u$ the diameters of the circles are the same. The points T and E represent the total and effective stresses on the same plane and clearly total and effective shear stresses are equal. Therefore, effective stresses are

$$\sigma' = \sigma - u \tag{6.6}$$

$$\tau' = \tau \tag{6.7}$$

From the definitions of the shear stress parameter q and the mean stress parameter p given in Chapter 2 and substituting $\sigma'_1 = \sigma_1 - u$, etc., it is easy to show that

$$p' = p - u \tag{6.8}$$

$$q' = q \tag{6.9}$$

From Eqs (6.7) and (6.9) total and effective shear stresses are identical and most authors use shear stresses without primes all the time. In my work and teaching, and in this book, I use τ' and q' when I am considering analyses in terms of effective stress and τ and q for total stresses. I know that this is strictly unnecessary but I find that the distinction between total and effective shear stresses is helpful, particularly for teaching.

6.5 IMPORTANCE OF EFFECTIVE STRESS

The principle of effective stress is absolutely fundamental to soil mechanics and its importance cannot be overstated. This is the way in which soil behaviour due to loading is related to behaviour due to changes of groundwater pressure.

Although most texts on soil mechanics examine the validity of the principle and the meaning of effective stress by considering the interparticle forces and the intergranular contact areas, there really is no need to do this and the necessary assumptions are not always supported by experimental evidence. Nevertheless, no conclusive evidence has yet been found that invalidates Terzaghi's original postulate, at least for saturated soils at normal levels of engineering stress, and the principle of effective stress is accepted as a basic axiom of soil mechanics.

Because total and effective normal stresses are different (except when pore pressures are zero) it is absolutely essential to distinguish between the two. The effective stresses σ' and τ' are always denoted by primes while the total stresses σ and τ do not have primes. Any equation should have all total stresses, or all effective stresses, or total and effective stresses should be related correctly by the pore pressure. Engineers doing design calculations (or students doing examination questions) should always be able to say whether they are dealing with total or effective stresses.

From Figs 6.1 and 6.2, and making use of Eqs (6.1) to (6.6), we can calculate the vertical effective stress σ'_z at any depth in the ground for any position of the groundwater. If you try some examples you will discover that if the water table is below the ground level the effective stress depends on the position of the water table. If, on the other hand, the ground level is submerged, as in the bed of a river, lake or sea, the effective stress is independent of the depth of water; this means that the effective stresses in soil in the bed of a duck pond will be the same as those in the bed of the deep ocean where the water depth may exceed 5 km. In doing these calculations remember that free water which can slosh around (i.e. in a river, lake or sea) will apply a total stress to the soil (and to dams and submarines), but water in the pores of the soil has a pore pressure; these water pressures need not always be equal.

Submarines and fish illustrate effective stresses. Sea water applies total stresses to the skin of both. In a submarine the internal (pore) pressure is zero (atmospheric) so the skin of the submarine must be very strong, but in a fish the pressures in the blood and in the soft tissues are very nearly equal to the external water pressure so the skin and skeleton of the fish can be very weak and soft. In both cases the stresses on the skins are equivalent to effective stresses in soils.

6.6 DEMONSTRATIONS OF EFFECTIVE STRESS

The effective stress equation (6.6) can be written in terms of changes Δ so that

$$\Delta\sigma' = \Delta\sigma - \Delta u \tag{6.10}$$

This shows that effective stresses may be changed—causing measurable effects—by changing either the total stress with the pore pressure constant or by changing the pore pressure with the total stress constant. Note also that if the total stress and the pore pressure are changed equally the effective stress remains constant and the soil state does not change.

Figure 6.5(a) illustrates settlements $\Delta\rho$ caused by loading a foundation by $\Delta\sigma$ while the pore pressures in the ground remain constant so that $\Delta\sigma' = \Delta\sigma$. Figure 6.5(b) illustrates settlements $\Delta\rho$ caused by extraction of groundwater. Pumping lowers the water table by Δh_w so that pore pressures reduce by $\Delta u = \gamma_w \Delta h_w$. From Eq. (6.10), with $\Delta\sigma = 0$, the reduction of pore pressure causes an increase of effective stress $\Delta\sigma'$. The principle of effective stress states that if the change of foundation loading $\Delta\sigma$ is the same as the change of pore pressure Δu due to lowering of groundwater the settlements will be the same. In other words, it is simply the change of effective stress that affects the soil behaviour.

A simple experiment which demonstrates the action of effective stresses is illustrated in Fig. 6.6. This shows the influence of pore pressure on the capacity of deep and shallow foundations. The soil should be fine to medium sand; if it is too coarse it will become unsaturated when the water table is lowered and if it is too fine pore pressures may not equalize in a reasonable time.

(a) Foundation loading

(b) Groundwater extraction

Figure 6.5 Settlements due to changing effective stresses.

Figure 6.6 Rising groundwater experiment.

Place the gravel and sand in water to ensure they are saturated and then open the valve to lower the water table to the gravel. Place a heavy foundation (a steel cylinder about 40 mm in diameter and 80 mm long works very well) and an eccentrically loaded pile as illustrated. Close the valve and raise the water table by pouring water into the standpipe; if the sand and gravel remained saturated it will only be necessary to fill the standpipe. As the water table and the pore pressures rise, effective stresses will fall and both foundations will fail.

Another simple demonstration of effective stresses is the stiffness and strength of a vacuum packed bag of coffee beans. As long as the vacuum is intact the bag is relatively stiff and strong because the negative pore pressures result in positive effective stresses. However, if you puncture the bag with a small pin prick it will become much less stiff and strong because the pore pressures rise so the effective stresses reduce. You can do the same experiment more cheaply using coarse sand or gravel in a self-sealing plastic bag.

6.7 VOLUME CHANGE AND DRAINAGE

As soil is loaded or unloaded due to changes of effective stress it will generally change in volume. However, because the soil grains themselves are very stiff the volume change of the grains is negligible and so the volume change of the soil must be due to rearrangement of the grains and

Figure 6.7 Volume changes in soil.

(a) Loose soil   (b) Dense soil

Figure 6.8 Drainage of clay beneath an embankment.

changes in volume of the voids. At small effective stress the spacing of the grains may be loose and at high stresses it will be dense, as shown in Fig. 6.7. If the pore pressure u_0 remains constant then the changes of total and effective stresses are the same ($\Delta\sigma' = \Delta\sigma$; see Eq. 6.10). If the volume of the soil grains remains constant then, in Fig. 6.7, the change of volume of the soil ΔV is the same as the volume of water expelled ΔV_w.

In saturated soil changes in volume must be due to seepage of water through the soil and so soil compression is rather like squeezing water from a sponge. In a laboratory, test water will seep to the boundaries of the sample while, in the ground, water will seep to the surface or to natural drainage layers in the soil. For example, Fig. 6.8 illustrates an embankment built on a bed of clay sandwiched between layers of sand which act as drains. As the embankment is constructed water will seep from the clay to the sand layers as indicated.

There must, of course, be sufficient time for the water to seep through the soil to permit the volume change to occur; otherwise the pore pressure will change. As a result there must be some relationship between the rate at which the loads are applied, the rate of drainage and the behaviour of the soil and pore pressure.

6.8 DRAINED LOADING, UNDRAINED LOADING AND CONSOLIDATION

The relative rates at which total stresses are applied and at which the seepage takes place are of critical importance in determining soil behaviour. The limiting conditions are illustrated in Figs 6.9 and 6.10.

Figure 6.9(a) illustrates an increment of total stress $\Delta\sigma$ applied slowly, over a long period of time. This could represent loading in a laboratory test or in the ground. If the loading is applied very slowly water will be able to seep from the soil as the total stresses increase. There will be no change of pore pressure, as shown in Fig. 6.9(c), and the volume changes will follow the change of loading, as shown in Fig. 6.9(b). Because the pore pressures remain constant at u_0, the changes of effective stress follow the change of total stress, as shown in Fig. 6.9(d). When the stresses remain constant at $\sigma_0' + \Delta\sigma'$, the volume remains constant at $V_0 - \Delta V$. This kind of

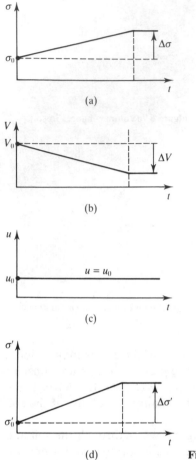

(a)

(b)

(c)

(d)

Figure 6.9 Characteristics of drained loading.

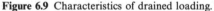

relatively slow loading is called drained because all the drainage of water takes place during the loading. The most important feature of drained loading is that the pore pressures remain constant at u_0, which is known as the steady state pore pressure.

Figure 6.10(a) illustrates the same increment of total stress $\Delta\sigma$ as in Fig. 6.9, but now applied so quickly that there was no time for any drainage at all and so the volume remains constant, as shown in Fig. 6.10(b). If the loading was isotropic with no shear distortion and undrained with no volume change then nothing has happened to the soil. From the principle of effective stress this means that the effective stress must remain constant, as shown in Fig. 6.10(d), and, from Eq. (6.10), the change in pore pressure is given by

$$\Delta\sigma' = \Delta\sigma - \Delta u = 0 \qquad (6.11)$$

$$\Delta u = \Delta\sigma \qquad (6.12)$$

This increase in pore pressure gives rises to an initial excess pore pressure $\bar{u}_i$, as shown in Fig. 6.10(c). Notice that the pore pressure u consists of the sum of the steady state pore pressure u_0 and the excess pore pressure $\bar{u}$; if the pore pressures are in equilibrium $u = u_0$ and $\bar{u} = 0$. Relatively quick loading is known as 'undrained loading' because there is no drainage of water during the loading. The most important feature of undrained loading is that there is no change of volume.

At the end of the undrained loading the pore pressure is $u = u_0 + \bar{u}_i$, where u_0 is the steady state, or equilibrium, pore pressure and $\bar{u}_i$ is an initial excess pore pressure. This excess pore

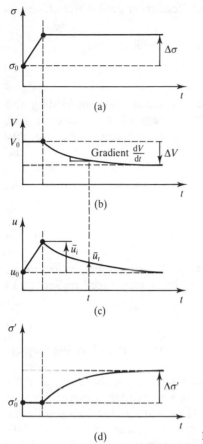

Figure 6.10 Characteristics of undrained loading and consolidation.

pressure will cause seepage to occur and, as time passes, there will be volume changes as shown in Fig. 6.10(b). The volume changes must be associated with changes of effective stress, as shown in Fig. 6.10(d), and these occur as a result of decreasing pore pressures, as shown in Fig. 6.10(c); at some time t the excess pore pressure is $\bar{u}_t$. The drainage of water is driven by the excess pore pressure and so, as the excess pore pressure decreases, the rate of volume change, given by the gradient dV/dt, also decreases, as shown in Fig. 6.10(b). Notice that while there are excess pore pressures in the soil, water pressures outside the surface of the soil will not be the same as the pore pressures; this means that the pore pressure in soil behind a new quay wall need not be the same as the pressure in the water in the dock.

This dissipation of excess pore pressure accompanied by drainage and volume changes is known as consolidation. The essential feature of consolidation is that there are excess pore pressures $\bar{u}$ that change with time. Usually, but not always, the total stresses remain constant. Consolidation is simply compression (i.e. change of volume due to change of effective stress) coupled with seepage. At the end of consolidation, when $\bar{u}_\infty = 0$ after a long time, the total and effective stresses and the volume are all the same as those at the end of the drained loading shown in Fig. 6.9. Thus, the changes of effective stress for undrained loading plus consolidation are the same as those for drained loading.

In the simple examples of drained and undrained loading illustrated in Figs 6.9 and 6.10, the increment of loading was positive so that the soil compressed as water was squeezed out. Exactly the same principles apply to unloading where the increment is negative and the soil swells as water is sucked in by the negative excess pore pressure. Readers should sketch diagrams like Figs 6.9 and 6.10 for an increment of unloading.

6.9 RATES OF LOADING AND DRAINAGE

When distinguishing between drained and undrained loading it is relative rates of loading and seepage that are important, not the absolute rate of loading. Seepage of water through soil, which will be covered in more detail in Chapter 17, is governed by the coefficient of permeability k. Figure 6.11 illustrates seepage with velocity V through an element of soil δs long. At one end there is a drain where the pore pressure is $u_0 = \gamma_w h_{w0}$ and at the other end there is an excess pore pressure given by $\bar{u} = \gamma_w \bar{h}_w$. The difference in the levels of water in the standpipes is $\delta h_w = \bar{h}_w$ and the hydraulic gradient is given by

$$i = \frac{\delta h_w}{\delta s} \tag{6.13}$$

(Hydraulic gradient should really be defined in terms of the hydraulic potential P instead of the head h_w, but if the flow is horizontal these are the same; potential is introduced in Sec. 17.3.) The basic rule for seepage is Darcy's law, given by

$$V = ki \tag{6.14}$$

where the coefficient of permeability k has the units of velocity. The value of k is the seepage velocity of water through soil with unit hydraulic gradient.

Values for the coefficient of permeability for soils depend largely on the grain size (or more particularly on the size of void spaces through which the seepage takes place). Typical values for k for different grain sizes are given in Table 6.1. (For some natural clay soils the value of k may be considerably less than 10^{-8} m/s.) Notice the very large range (more than $\times 10^6$) of permeability for typical soils. Under a unit hydraulic gradient, water will travel 1 m through gravel in less than $10^2 = 100$ s and 1 m through clay in more than 10^8 s, which is about 3 years.

In civil engineering and related activities loads are applied to the ground at different rates and some typical examples are given in Table 6.2. Again, notice the very large range (more than $\times 10^9$) in the durations, or rates, of loading or unloading in these examples.

In any geotechnical calculation or analysis it is absolutely essentially to state whether the calculation is for drained or undrained loading, and we will discover that different analyses are required for each in later chapters. What is important is the relative rates of loading and drainage—is there enough time during the loading to allow drainage to occur or is the loading so fast that there will be no drainage? Of course, in reality, neither condition will be satisfied

Figure 6.11 Seepage of water through soil.

Table 6.1 Values of coefficient of permeability of soils

Grain size	k (m/s)
Gravel	$> 10^{-2}$
Sand	$10^{-2} - 10^{-5}$
Silt	$10^{-5} - 10^{-8}$
Clay	$< 10^{-8}$

Table 6.2 Durations of typical engineering constructions

Event	Duration
Shock (earthquake, pile driving)	< 1 s
Ocean wave	10 s
Excavate trench	10^4 s ≈ 3 h
Load small foundation	10^6 s ≈ 10 days
Large excavation	10^7 s ≈ 3 months
Embankment dam	10^8 s ≈ 3 years
Natural erosion	10^9 s ≈ 30 years

absolutely and decisions must be made as to whether the construction is more nearly drained or undrained.

Many engineers will assume that loading or unloading of a coarse-grained soil will be drained and of a fine-grained soil will be undrained. These assumptions are adequate for loading rates which are not at the extremes of those in Table 6.2. Very rapid loading of coarse-grained soil is likely to be undrained. Thus earthquakes, pile driving and ocean waves may generate excess pore pressures in sands which can cause liquefaction failures and which explain the change of pile capacity after a delay in driving. Very slow loading of clay slopes due to natural erosion is likely to be drained and pore pressures and slope angles of many natural clay slopes correspond closely to the fully drained, steady state conditions.

6.10 SUMMARY

1. In soils total stresses arise from the weight of the soil (including the soil grains and the pore water) and any other external loads from foundations, walls and free water. There are also pore pressures in the water in the voids.
2. The stresses that govern soil behaviour are effective stresses given by $\tau' = \tau$ and $\sigma' = \sigma - u$. As a result soils are affected equally by changes in total stress and pore pressure.
3. Volume changes in soil can only occur as water seeps through the pores and the rate of seepage is governed by the coefficient of permeability k. If soil is loaded slowly, compared with the rate of drainage, the pore pressures remain constant and volume changes occur during the loading which is called drained.
4. If soil is loaded quickly, compared with the rate of drainage, the volume remains constant, excess pore pressures arise and the loading is called undrained. Subsequently, consolidation occurs as the excess pore pressures dissipate and water seeps from the soil, causing volume changes.

WORKED EXAMPLES

Example 6.1: Calculation of vertical stress For the measuring cylinder of sand described in Example 5.4 (see Fig. 5.8), the total vertical stress, the pore pressure and the effective vertical stress at the base of the cylinder are:

(a) When the sand is loose and dry:

$$z = 0.40 \text{ m}$$
$$\gamma_d = 15.5 \text{ kN/m}^3$$
$$u = 0$$
$$\sigma_z = \gamma_d z = 15.5 \times 0.40 = 6.2 \text{ kPa}$$

(b) When the sand is loose and saturated:

$$z = 0.40 \text{ m}$$
$$\gamma = 19.5 \text{ kN/m}^3$$
$$\sigma_z = \gamma z = 19.5 \times 0.4 = 7.8 \text{ kPa}$$
$$u = \gamma_w h_w = 9.81 \times 0.4 = 3.9 \text{ kPa}$$
$$\sigma_z' = \sigma_z - u = 7.8 - 3.9 = 3.9 \text{ kPa}$$

(c) When the sand is dense and saturated:

$$z = 0.37 \text{ m}$$
$$\gamma = 20.4 \text{ kN/m}^3$$
$$z_w = 0.03 \text{ m}$$
$$\sigma_z = \gamma z + \gamma_z z_w = (20.4 \times 0.37) + (9.81 \times 0.03) = 7.8 \text{ kPa}$$
$$u = \gamma_w h_w = 9.81 \times 0.40 = 3.9 \text{ kPa}$$
$$\sigma_z' = \sigma - u = 7.8 - 3.9 = 3.9 \text{ kPa}$$

Notice that densification of the soil by tapping the side of the cylinder did not change the total or effective stresses at the base of the cylinder. This is simply because the total weights of soil and water in the cylinder did not change.

Example 6.2: Calculation of stress in the ground The deep clay deposit in Fig. 6.12 has unit weight $\gamma = 20 \text{ kN/m}^3$ and the soil remains saturated even if the pore pressures become negative. For the groundwater conditions, (a) water table 6 m below ground level and (b) with water to a depth of 3 m above ground level, the vertical effective stresses at a depth of 3 m are:

(a) Water table at 6 m below ground level:

$$\sigma_z = \gamma z = 20 \times 3 = 60 \text{ kPa}$$
$$u = \gamma_w h_w = 10 \times -3 = -30 \text{ kPa}$$
$$\sigma_z' = \sigma_z - u = 60 + 30 = 90 \text{ kPa}$$

(a)   (b)   **Figure 6.12**

(b) Water surface 3 m above ground level:

$$\sigma_z = \gamma z + \gamma_w z_w = (20 \times 3) + (10 \times 3) = 90\,\text{kPa}$$

$$u = \gamma_w h_w = 10 \times 6 = 60\,\text{kPa}$$

$$\sigma_z' = \sigma_z - u = 90 - 60 = 30\,\text{kPa}$$

Example 6.3: Calculation of stress in the ground below a foundation   The concrete bridge pier in Fig. 6.13 is 4 m tall, it has an area of 10 m² and carries a load of 1 MN. (The unit weight of concrete is $\gamma_c = 20\,\text{kN/m}^3$.) The pier is founded on the bed of a tidal river where there is at least 5 m of sand with a unit weight of 20 kN/m³. The river bed is at low tide level and at high tide there is 3 m depth of water.

The total contact stress q between the soil and the base of the pier (i.e. the bearing pressure) arises from the weight of the concrete and the applied load and is

$$q = \gamma_c H_c + \frac{F}{A} = (4 \times 20) + \frac{1 \times 10^3}{10} = 180\,\text{kPa}$$

(a) At low tide:

$$\sigma_z = \gamma z + q = (20 \times 2) + 180 = 220\,\text{kPa}$$

$$u = \gamma_w h_w = 2 \times 10 = 20\,\text{kPa}$$

$$\sigma_z' = \sigma_z - u = 220 - 20 = 200\,\text{kPa}$$

(a)   (b)   **Figure 6.13**

(b) At high tide (note that q is reduced by uplift from the water pressure below the foundation):

$$\sigma_z = \gamma z + \gamma_w z_w + (q - \gamma_w z_w)$$
$$= (20 \times 2) + (10 \times 3) + (180 - (10 \times 3)) = 220 \, \text{kPa}$$
$$u = \gamma_w h_w = 10 \times 5 = 50 \, \text{kPa}$$
$$\sigma'_z = \sigma_z - u = 220 - 50 = 170 \, \text{kPa}$$

Notice that in this case the increase of the water depth has reduced the effective stress in the ground; this is because of a reduction of the bearing pressure due to uplift.

Example 6.4: Calculation of stress below an embankment The soil profile in Fig. 6.14 consists of 4 m clay over 2 m sand over rock: the unit weights of all the natural materials are 20 kN/m³ and the steady state water table is at ground level. A wide embankment 4 m high is constructed from fill with a unit weight of 15 kN/m³. The total and effective vertical stresses at the centre of the clay and at the centre of the sand (a) before the embankment is constructed, (b) immediately after it is completed and (c) after a very long time are:

(a) Before construction of the embankment:

- in the clay:

$$\sigma_z = \gamma z = 20 \times 2 = 40 \, \text{kPa}$$
$$u = \gamma_w h_w = 10 \times 2 = 20 \, \text{kPa}$$
$$\sigma'_z = \sigma_z - u = 40 - 20 = 20 \, \text{kPa}$$

- in the sand:

$$\sigma_z = \gamma_z = 20 \times 5 = 100 \, \text{kPa}$$
$$u = \gamma_w h_w = 10 \times 5 = 50 \, \text{kPa}$$
$$\sigma'_z = \sigma_z - u = 100 - 50 = 50 \, \text{kPa}$$

(b) Immediately after construction of the embankment the sand is drained so the pore pressure remains constant. The embankment is wide so there are no horizontal strains, the clay is undrained and the effective stresses remain unchanged:

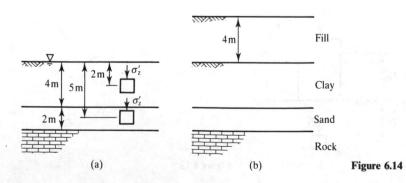

(a) (b) **Figure 6.14**

- in the clay:

$$\sigma_z = \sum \gamma z = (4 \times 15) + (2 \times 20) = 100 \text{ kPa}$$

$$\sigma'_z = 20 \text{ kPa, as in (a)}$$

$$u = \sigma_z - \sigma'_z = 100 - 20 = 80 \text{ kPa}$$

- in the sand:

$$\sigma_z = \sum \gamma z = (4 \times 15) + (20 \times 5) = 160 \text{ kPa}$$

$$u = 50 \text{ kPa}$$

$$\sigma'_z = \sigma_z - u = 160 - 50 = 110 \text{ kPa}$$

(c) After a very long time the excess pore pressures in the clay will have dissipated to the steady state conditions corresponding to the water table at original ground level:

- in the clay:

$$\sigma_z = 100 \text{ kPa, as in (b)}$$

$$u = 20 \text{ kPa, as in (a)}$$

$$\sigma'_z = \sigma_z - u = 100 - 20 = 80 \text{ kPa}$$

- in the sand there has been no change of total stress or pore pressure and the stresses are the same as those in (b).

REFERENCE

Terzaghi, K. (1936) 'The shearing resistance of saturated soil and the angle between the planes of shear', *Proceedings of 1st International SMFE Conference, Harvard, Mass.*, Vol. 1, pp. 54–56.

CHAPTER SEVEN

LABORATORY TESTING OF SOILS

7.1 PURPOSES OF LABORATORY TESTS

Testing soil samples in the laboratory plays an important role in soil mechanics research and civil engineering practice. Almost all we know about soil behaviour has been learned from laboratory tests. Tests may be carried out on small samples of soil to examine the characteristics of the soil or on models of soil structures to examine how slopes, walls and foundations deform and collapse. In this chapter we will consider tests on soil samples. Laboratory tests are carried out for a number of purposes, the most important being:

1. For description and classification of a particular soil (see Chapter 5).
2. To investigate the basic mechanical behaviour of soils and to develop theories for soil behaviour (see Chapters 8 to 12).
3. To determine design parameters (i.e. numerical values for strength, stiffness and permeability) for geotechnical analyses (see Chapters 17 to 23).

Laboratory tests may be carried out on samples that are intact or have been completely reconstituted. In reconstituted samples the soil has been mixed at a relatively large water content and then recompressed (see Chapter 8). In this way any structure developed in the soil in the ground due to deposition or ageing is removed and the tests measure the fundamental behaviour and intrinsic properties of the soil. Intact samples are recovered from the ground with minimum disturbance (see Chapter 16); thus they contain the *in situ* structure and retain the properties of the soil in the ground.

Most of the analyses of geotechnical structures described in Chapters 18 to 23 and used for routine design were developed for simple soils which behave more or less like the theories described in Chapters 8 to 12. These analyses may not be applicable to soils whose behaviour differs significantly from these simple theories, in which case special methods will be required which are outside the scope of this book. An important and often neglected purpose of soil testing is to examine soil for unexpected or strange behaviour. This is best done by comparing the behaviour of intact samples with the basic theories and with the behaviour of the same soil reconstituted and recompressed to the same state.

7.2 STANDARD TESTS AND SPECIFICATIONS

Many of the routine soil tests are very carefully and precisely specified in a number of national standards and codes of practice. In the United Kingdom the standard is BS 1377:1991, *Methods of Test for Soils for Civil Engineering Purposes*, and similar standards exist in other countries and regions. You should certainly look at a copy of the standards for soil testing relevant to your region to see exactly what they cover. Most of these standards follow what might be called a cookery book method: you do this, you do that and you serve up the result in this or in that way. There are, however, difficulties with the cookery book approach for soil testing which arise from the characteristics of soil strength and stiffness described in Chapters 8 to 13.

The values obtained from a particular test will obviously depend to a greater or lesser extent on details of the equipment and procedures used and for some tests, particularly those that measure the nature and state of a soil, it is essential that the tests follow standard procedures. This is because the parameters being measured (e.g. grading and Atterberg limits) are intrinsic properties of the material and different laboratories and different workers should obtain identical results for the same soil.

While it is possible and desirable to set standards for construction of equipment and for calibration of instruments to ensure that the accuracy of the observations is acceptable (or at least known), it is not so easy to specify tests that measure soil strength and stiffness because of the many important factors that affect these parameters. Instead, engineers should determine what parameters are required for a particular analysis, determine what factors will influence these within the theories described in Chapters 8 to 13 and then devise tests that take account of these. The engineer will need to specify not only the loading path applied in the test but also, equally importantly, the loads applied to the sample before the test starts.

I am not going to describe the standard equipment and soil tests in detail. Most of the standard apparatus and routine tests are described at length in a three-volume book by Head (1980, 1982 and 1986) and in various standards and codes of practice. All engineers concerned with groundworks should carry out simple classification, consolidation, shear and triaxial tests for themselves at least once in their career; they should also carry out simple foundation, slope stability and retaining wall experiments. The emphasis of this work should be on handling equipment and soil samples, good scientific practice and analysis and interpretation of test results within simple theories. They should also play around with soils and soil-like materials at home, in their garden and at the beach.

7.3 BASIC CLASSIFICATION TESTS

As discussed in Chapter 5, the nature of a soil is described principally by the grading (i.e. the distribution of particle sizes) and the mineralogy, while the state is described by the current water content and unit weight (together with the current stresses).

(a) Measurement of Grading

The distribution of particle sizes in a soil is found by sieving and sedimentation. Soil is first passed through a set of sieves with decreasing aperture size and the weight retained on each sieve recorded. The smallest practical sieve has an aperture size of about 0.07 mm, corresponding roughly to the division between silt and sand. Silt-sized particles can be separated by sedimentation making use of Stoke's law, which relates the settling velocity of a sphere to its diameter.

A rapid estimate of grading can be made by sedimentation in a jam jar or milk bottle. Take a sample about one-third of the height of the container, fill the container with water and shake it up. Quickly stand the jar or bottle upright and leave it for several hours. You can see and estimate the grading of gravel, sand and silt; clay will remain in suspension for a long time and any material floating on the surface is likely to be organic (i.e. peat). This is a test frequently used by gardeners.

(b) Measurement of Water Content and Unit Weight

The water content of a soil is defined as

$$w = \frac{W_w}{W_s} \tag{7.1}$$

and the unit weight γ is defined as

$$\gamma = \frac{W}{V} \tag{7.2}$$

where W_w is the weight of water evaporated by heating soil to 105°C until the weight is constant, W_s is the weight of dry soil and $W = W_w + W_s$ is the weight of a sample with volume V. These weights can be measured by simple weighing and the volume of a cylindrical or cubic sample determined by direct measurement.

(c) Measurement of Atterberg Limits

For coarse-grained soils the engineering properties are governed largely by the grading and, to a lesser extent, by the shape, texture and mineralogy of the grains, but the properties of fine-grained clay soils depend largely on the type of clay. The basic behaviour of plastic clay soils can be assessed from the Atterberg limits (i.e. liquid limit, plastic limit and plasticity index) described in Sec. 5.6. The liquid limit determines the water content at which the soil has weakened so much that it starts to flow like a liquid. The plastic limit determines the water content at which the soil has become so brittle that it crumbles.

Liquid limit tests The two alternative liquid limit tests are illustrated in Fig. 7.1. In the Casagrande test in Fig. 7.1(a) a small slope is failed by bumping a dish on to a rubber block.

(a) Casagrande liquid limit test

(b) Fall cone liquid limit test

(c) Plastic limit test

Figure 7.1 The Atterberg limits tests.

In the fall cone test a small cone-shaped foundation penetrates the soil. The precise details of the geometries, weights and so on are arranged so that the soil has a strength of approximately 1.5 kPa when it is at the liquid limit. In each case the sample has a high water content and is soft enough to be moulded into the container using a knife or spatula. The tests are repeated with slightly different water contents until the precise requirements of the tests are met.

Plastic limit test The test consists of rolling a 3 mm diameter thread of soil while the water evaporates and the water content decreases until the thread splits and crumbles. The failure of the thread corresponds to a strength of approximately 150 kPa. Notice that a strength of 150 kPa corresponds to the division between stiff and very stiff clay in Table 5.2. Remember the plasticity index PI given by

$$PI = LL - PL \tag{7.3}$$

This is an important intrinsic soil parameter. Because the Atterberg limits determine the conditions of soil at certain well-specified strengths, the results can be used to estimate a number of other important intrinsic soil properties, as discussed in Chapter 9. Further discussion of the Atterberg limits is given in Sec. 5.6.

7.4 MEASUREMENT OF COEFFICIENT OF PERMEABILITY

Seepage of water through soil, discussed in Chapter 17, is governed by Darcy's law:

$$V = ki \tag{7.4}$$

where k, the coefficient of permeability, is a soil parameter. The value of k depends principally on the grain size and specific volume (or more properly on the void size, which is related to the grain size and specific volume). Permeability can be measured in laboratory tests in a constant head permeameter, for soil with relatively large permeability, or in a falling head permeameter, for soils with relatively low permeability; these are illustrated in Fig. 7.2. In both cases water flows through a soil sample and the rates of flow and the hydraulic gradients are measured.

(a) Constant Head Permeability Tests

In the constant head test illustrated in Fig. 7.2(a) water from a constant head tank flows through the sample in a cylinder and is collected in a measuring jar. Two standpipes measure the pore pressure and potential (see Sec. 17.3) at two points as shown. The flow is steady state and, from the observations,

$$V = \frac{\Delta Q}{A \Delta t} \tag{7.5}$$

$$i = \frac{\Delta P}{\Delta s} \tag{7.6}$$

and hence a value for k can be determined. In practice it is best to vary the rate of flow in stages and plot V against i; in this way you can verify Darcy's law and evaluate k.

(b) Falling head test

Figure 7.2 Permeameter tests.

(b) Falling Head Permeability Tests

In the falling head test illustrated in Fig. 7.2(b) water flows through the sample as the level of water in the standpipe drops. Over a time interval δt the rate of flow is

$$q = -a \frac{\delta P}{\delta t} = Ak \frac{P}{L} \qquad (7.7)$$

and hence, in the limit,

$$-\frac{dP}{P} = \frac{Ak}{aL} dt \qquad (7.8)$$

Integrating with the limits $P = P_0$ at $t = 0$ we have

$$\ln\left(\frac{P_0}{P}\right) = \frac{Ak}{aL} t \qquad (7.9)$$

and you can determine a value for k by plotting $\ln(P_0/P)$ against t and finding the gradient. Notice that in a falling head test the effective stresses change because the pore pressures change as the level of water in the standpipe falls. Any volume changes that occur as a result of these changes of effective stress have to be neglected.

Values of the coefficient of permeability measured in laboratory permeameter tests are often highly inaccurate, for a variety of reasons such as anisotropy (i.e. values of k different for horizontal and vertical flow) and small samples being unrepresentative of large volumes of soil in the ground, and in practice values of k measured from *in situ* tests are much better.

7.5 PRINCIPAL FEATURES OF SOIL LOADING TESTS

Soil strength and stiffness are investigated and measured in tests in which soil samples are loaded and unloaded and the resulting stresses and strains are measured. The requirements for testing soils are rather like those for testing metals, concrete and plastics, but the special feature of soil strength and stiffness impose special requirements. The most important of these are:

1. Total stresses and pore pressures must be controlled and measured separately so effective stresses, which govern soil behaviour (see Secs 6.4 to 6.6) can be determined.
2. Drainage of water into, or out of, the sample must be controlled so that tests may be either drained (i.e. constant pore pressure) or undrained (i.e. constant volume) (see Secs 6.7 to 6.9).
3. To investigate soil stiffness, measurements must be made of small strains (see Chapter 13), but to investigate soil strength it is necessary to apply large strains, sometimes greater than 20 per cent.
4. Because soils are essentially frictional it is necessary to apply both normal and shear stresses. This can be done either by applying confining pressures to cylindrical or cubic samples or by applying normal stresses in direct shear tests (see Fig. 3.3); the relationships between the principal stresses on cylindrical samples and the normal and shear stresses on shear samples were discussed in Sec. 3.2.

During a test the total stresses could be changed, or held constant, and the resulting strains measured; such a test is called stress controlled. Alternatively, the strains could be changed, or held constant, and the resulting stresses measured; such a test is called strain controlled. In a particular test one set of stresses (i.e. axial or vertical) could be stress controlled and another set (i.e. radial or horizontal) could be strain controlled or vice versa.

Loads may be applied to soil samples by rigid plates or by fluid pressures acting on flexible membranes. In the first case the displacements and strains are uniform but the stresses may vary across the plate; in the second case the stresses will be uniform but the strains may vary. Rigid plates may be smooth, in which case shear stresses should be zero and so the faces of the sample are principle planes or they may be rough, in which case there will be both shear and normal stresses to be measured.

To control drainage and measure pore pressures the sample must be isolated within an impermeable membrane and the pore water connected through drainage leads to a pressure transducer and volume gauge, as shown in Fig. 7.3. (This shows details of drainage connections in a typical triaxial test apparatus but the general principles apply also to other soil testing apparatus.) There is a second drainage lead to the sample with a flushing valve. This is to allow water to be flushed through the drainage leads and the bottom drain for de-airing; this

Figure 7.3 Control of drainage and measurement of pore pressure in soil tests.

is an important requirement of soil testing. If both valves are closed the sample is undrained and if the drainage valve is open the sample is drained; the flushing valve is normally closed and it is only opened when the drainage leads are being flushed. The back pressure u_0 may be atmospheric or at some elevated pressure. Sometimes special tests are carried out in which the pore pressures are changed independently of the total stresses.

The general requirements of soil tests described above are often conflicting and a number of different soil tests have been developed for different specific purposes. The principal tests in routine use in practice are the oedometer test, the direct shear test and the triaxial test, which will now be described. If you read the literature of soil mechanics and become sufficiently interested to specialize in this area you will come across many other special tests; all you have to do is work out what are the boundary conditions and the abilities and limitation of the tests.

7.6 ONE-DIMENSIONAL COMPRESSION AND CONSOLIDATION (OEDOMETER) TESTS

One of the simplest forms of soil loading test is the one-dimensional oedometer test illustrated in Fig. 7.4. The soil sample is a disc contained in a stiff metal cylinder so that radial strains are zero. Porous discs at the top and bottom act as drains and so seepage of pore water is vertical and one-dimensional.

In the conventional apparatus illustrated in Fig. 7.4(a) the axial stress σ_a is applied by adding (or removing) weights so the loading is stress controlled and applied in stages. The axial strain ε_a is measured using a displacement transducer or a dial gauge. The pore pressures in the top drain u_t are zero. The pore pressures in the bottom drain u_b are usually zero but in some special oedometers the bottom drain may be closed and values of u_b measured.

In the Rowe cell illustrated in Fig. 7.4(b) the axial stress σ_a is applied by fluid pressure in a rubber diaphragm so the loading is stress controlled and may be either applied in stages or varied smoothly in continuous loading tests. The axial strain ε_a is measured using a displacement transducer mounted on the stiff top drainage lead. The top and bottom drains are connected to drainage apparatus like that illustrated in Fig. 7.3 so that either or both top and bottom faces of the sample may be drained (i.e. constant pore pressure) or undrained (i.e. the drainage valve is closed).

Oedometer tests may be used to investigate compression and swelling of soil (i.e. the relationship between effective stress and volumetric strain) or consolidation (i.e. the relationship between compression and seepage). Remember the distinctions between drained loading, undrained loading and consolidation discussed in Secs 6.8 and 6.9. One-dimensional compression

(a) Conventional oedometer

(b) Rowe cell

Figure 7.4 One-dimensional consolidation (oedometer) tests.

and swelling of soil is discussed in Sec. 8.4 and one-dimensional consolidation is discussed in Chapter 14.

7.7 SHEAR TESTS

The two forms of shear test used for soil testing are illustrated in Fig. 7.5. In the direct shear box test illustrated in Fig. 7.5(a) the sample is in a split box and is obliged to shear along the horizontal plane defined by the halves of the box. The normal stress σ_n is applied by weights and the shear stress τ_n is usually applied at a constant rate of displacement. The vertical and horizontal displacements δn and δh are measured using dislacement transducers or dial gauges. Drains are provided at the top and bottom and the pore pressures u_t and u_b are zero. Tests on clays could be undrained if they were carried out quickly, so there was negligible drainage during the test, but as the pore pressures in the sample are not measured effective stresses are unknown. It is fairly obvious looking at Fig. 7.5(a) that the states of stress and strain within the sample are likely to be non-uniform, particularly near the ends of the box and at relatively large strains.

The design of the simple shear apparatus avoids non-uniform strains by allowing the sides to rotate. The most common type, known as the NGI (Norwegian Geotechnical Institute) simple shear apparatus, is illustrated in Fig. 7.5(b). The sample is cylindrical and is sealed inside a rubber sleeve like a triaxial sample (see Sec. 7.8). The rubber sleeve has a spiral wire reinforcement which prevents radial strains but permits shear strains as shown. Applications of the normal and shear stresses and measurements of strains are generally similar to those used for direct shear tests. The drain at the bottom is connected to drainage apparatus like that shown in Fig. 7.3, so that tests may be drained or undrained with measurements of pore pressure.

Notice that if the shear stresses and horizontal displacements in the shear tests in Fig. 7.5 are zero, the conditions are just the same as those in the one-dimensional compression tests in Fig. 7.4.

A major problem with direct and simple shear tests arises with interpretation of the test results. In the apparatus illustrated in Fig. 7.5 only the shear stresses τ_n and σ_n on horizontal planes are measured and the stresses on the vertical planes τ_h and σ_h in Fig. 7.6(a) are unknown. This means that we can only plot one point T on the Mohr diagram shown in Fig. 7.6(b). There are many Mohr circles that pass through the point T; two possibilities are shown. In some special simple shear test apparatus the stresses τ_h and σ_h on the vertical planes are measured, and in this case the Mohr circle is properly defined, but for the conventional tests in Fig. 7.5 it is not certain that the stresses measured, τ_n and σ_n, are those on the most critical planes.

(a) Direct shear   (b) Simple shear

Figure 7.5 Shear tests.

(a)   (b)   **Figure 7.6** Interpretation of shear tests.

7.8 CONVENTIONAL TRIAXIAL COMPRESSION TESTS

The trixial test is by far the most common and versatile test for soils. The conventional apparatus and the standard test procedures were described in detail by Bishop and Henkel (1962) in their standard text. Most of these are still widely used today, although many of the instruments have been superseded by modern electronic devices.

The basic features of the conventional triaxial tests are shown in Fig. 7.7. The soil sample is a cylinder with height about twice the diameter; sizes commonly used in the United Kingdom are 38 and 100 mm diameters (originally $1\frac{1}{2}$ and 4 in). The sample is enclosed in a thin rubber sleeve sealed to the top platen and to the base pedestal by rubber O-rings. This is contained in a water-filled cell with a cell pressure σ_c. A frictionless ram passes through the top of the cell and applies a force F_a to the top platen; this is measured by a proving ring or by a load cell either inside or outside the cell, as shown. Axial displacements are measured by a displacement transducer attached to the loading ram. The cell and sample assembly are placed inside a loading frame and a motor drive applies a constant rate of strain loading. There is a drain at the base of the sample connected to flushing and drainage apparatus like that shown in Fig. 7.3; if the drainage valve is open the sample is drained and if it is closed the sample is undrained. Radial

Figure 7.7 Conventional triaxial apparatus.

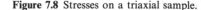

Figure 7.8 Stresses on a triaxial sample.

strains are not generally measured directly but are calculated from measurements of the axial and volumetric strains.

The axial and radial total stresses on the sample, σ_a and σ_r, are shown in Fig. 7.8(a). The radial stress is

$$\sigma_r = \sigma_c \tag{7.10}$$

where σ_c is the cell pressure as shown in Fig. 7.8(b) but σ_c acts also on the top of the sample. From Fig. 7.8 the axial stress σ_a is given by

$$\sigma_a = \sigma_r + \frac{F_a}{A} \tag{7.11}$$

or

$$\frac{F_a}{A} = \sigma_a - \sigma_r = \sigma'_a - \sigma'_r \tag{7.12}$$

If you go back to Sec. 3.2 you will see that F_a/A is the same as the deviator stress q. A simple way to think of the stresses in a triaxial sample is to decompose σ_a and σ_r into an isotropic state $\sigma_a = \sigma_r = \sigma_c$ as in Fig. 7.8(b) plus a deviatoric state $q = F_a/A$ as in Fig. 7.8(c); thus the force in the ram F_a (divided by the area of the sample) applies a stress that deviates from an isotropic state. Note that A is the current area of the sample allowing for changes of axial and volumetric strain. If the loading ram is raised away from the top platen so that $F_a = 0$ the state of stress is isotropic, with $\sigma_a = \sigma_r$. Isotropic compression and swelling of soil is discussed in Secs 8.2 and 8.3.

In a conventional triaxial test the sample would be isotropically compressed, either drained or undrained to the required initial state. The loading ram would then be lowered to touch the top platen, the axial strain set to zero and the sample sheared by increasing the deviator stress q, either drained or undrained, at a constant rate of strain. If the cell pressure σ_c is zero (in this case you need not fill the cell with water) the test is known as unconfined compression. There are a number of other special tests that can be carried out in the triaxial apparatus. These require special modifications to be made to the conventional apparatus, which are discussed in Sec. 7.9.

7.9 HYDRAULIC TRIAXIAL CELLS—STRESS PATH TESTS

Later we will discover that many features of soil strength and stiffness are governed by the initial state of the soil, its history of loading and unloading and the changes of axial and radial stress during loading or unloading. Consequently, in order to examine soil behaviour properly we will need to be able to control the axial and radial stresses, and perhaps the pore pressures, independently. In the conventional triaxial apparatus shown in Fig. 7.7 the axial stress is applied by strain-controlled loading and it is difficult to vary the axial stress in a controlled way.

Tests in which the paths of the effective stresses (i.e. the graph of σ'_a against σ'_r or the graph of q' against p') are varied, are called stress path tests and are carried out in hydraulic triaxial

(a) Conventional apparatus
   with hydraulic cylinder

(b) Special hydraulic
   triaxial cell

Figure 7.9 Hydraulic triaxial apparatus.

cells, illustrated in Fig. 7.9. Details of the sample, platens and drainage arrangements are the same as those for the conventional triaxial cell shown in Fig. 7.7, the principal difference being in the application of the axial stress. Another difference to notice is that the loading ram should be connected to the top platen so that extension tests can be carried out where $\sigma'_a < \sigma'_r$ and the force in the ram F_a is negative. (Note that σ'_a and σ'_r are always positive because uncemented soils cannot sustain tensile stresses and, in any case, the platens are not generally attached to the sample.)

A simple hydraulic triaxial cell can be made by adding a hydraulic cylinder to the loading ram, as illustrated in Fig. 7.9(a). Alternatively, special hydraulic triaxial cells are widely used in which a frictionless hydraulic ram is incorporated into the base of the cell, as illustrated in Fig. 7.9(b). In both cases the axial forces F_a should be measured independently using a load cell because it is inaccurate to calculate the value from measurements of the pressures in the hydraulic rams. Conventional strain-controlled triaxial tests can be carried out in both cells, in the first case by locking the hydraulic cylinder and using the motor drive in the loading frame as in a conventional test or, in the second case, by pumping fluid into the hydraulic ram at a constant rate from a screw ram.

In many modern hydraulic triaxial cells all the instruments are electronic and readings are made on a logger controlled by a microcomputer and the pressures in the axial ram, in the cell and in the pore pressure leads are applied through electronic pressure converters. In this case the microcomputer can be used to control the test and to record the results. Details of this equipment are beyond the scope of this book.

With a hydraulic triaxial cell like those shown in Fig. 7.9 the axial and radial stresses or strains and the pore pressure or volumetric strains can be changed independently. You can illustrate the test path by plotting total and effective stress paths using the axes σ_a vs σ_r and σ'_a vs σ'_r. However, because we are interested in shear and volumetric effects in soil behaviour it is more illustrative to plot stress paths using the axes q and p (or q' and p'). From Eqs (3.5) and (3.6), changes of total stress are given by

$$\delta q = \delta\sigma_a - \delta\sigma_r \tag{7.13}$$

$$\delta p = \tfrac{1}{3}(\delta\sigma_a + 2\delta\sigma_r) \tag{7.14}$$

and, from Eqs (6.8) and (6.9),

$$\delta q' = \delta q \tag{7.15}$$

$$\delta p' = \delta p - \delta u \tag{7.16}$$

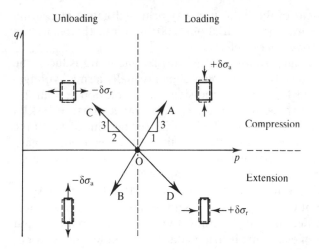

Figure 7.10 Typical stress paths available in hydraulic triaxial tests.

Hence, if you know $\delta\sigma_a$, $\delta\sigma_r$ and δu, you can easily plot stress paths using the axes q vs p and q' vs p'.

Figure 7.10 illustrates four simple total stress paths and also defines terms like compression, extension, loading and unloading. Note that in a triaxial apparatus σ_a and σ_r must always be positive; however, we can have $\sigma_a < \sigma_r$ (provided that the loading ram is attached to the top platen) and so q and q' can be positive or negative.

In Fig. 7.10 the four total stress paths correspond to increasing or decreasing either σ_a or σ_r while the other is held constant. Using Eqs (7.13) and (7.14) with either $\delta\sigma_r = 0$ or $\delta\sigma_a = 0$, you should show that the gradients dq/dp are 3 or $-\frac{3}{2}$. In Fig. 7.10 a distinction is made between loading or unloading (corresponding to increasing or decreasing p) and compression or extension (corresponding to positive or negative values of q). Notice that for compression the sample becomes shorter and fatter and for extension it becomes longer and thinner; the path OA corresponds to a conventional triaxial test with constant cell pressure, while path OD is like squeezing a toothpaste tube. During drained tests where the pore pressure u remains constant the total and effective stress paths are parallel, but during undrained tests in which the pore pressure generally changes, the total and effective stress paths are different.

7.10 COMMENTS ON SOIL TESTING

Although the routine soil tests described in this chapter are relatively simple there is a lot that can, and often does, go wrong with soil tests. Probably the most significant sources of error in measurements of soil parameters and behaviour in laboratory tests are:

1. Malfunctions and errors in the apparatus and in the instruments.
2. Incorrect detailed procedures in performing the tests.
3. Doing the wrong test or measuring the wrong parameter for a particular application.

The last of these is simply a matter of sound understanding of the basic theories involved, rather than blindly following a cookery book approach. The purpose of this book is to develop this sound understanding. The first two are largely a matter of care and attention and experience. In assessing the quality of a set of test results it is essential to distinguish very carefully and clearly between the accuracy and the resolution of the instruments. The resolution (or precision) of an observation is the smallest increment that can be discerned, while the accuracy is the limit within which you can be absolutely confident of the data. For a typical dial gauge measuring small displacements, the resolution and accuracy are both about 0.001 mm, but the resolution and accuracy of electronic instruments are often very different.

For a typical electronic load cell, pressure transducer or displacement transducer the resolution is linked to the electronics which converts an analogue signal (usually a small voltage) to a digital signal. For a 16-bit converter, using 1 bit for the sign, the resolution is 1 in 2^{15} ($\approx 30\,000$) of the full-scale reading, so for a pore pressure transducer with a range of 0 to 1000 kPa the resolution is about 0.03 kPa. The accuracy depends on the linearity (or non-linearity) of the calibration constant between pressure and voltage and on the stability of the electronic signals. With most instruments commonly used in soil testing you will be doing well to achieve an accuracy better than ± 1 kPa, which is very different from the resolution.

Another factor is in detection of malfunctions in instruments. It is usually fairly easy to see whether a dial gauge or proving ring is not working properly, but it is much less easy to detect malfunctions in electronic instruments provided that they continue to produce reasonable output signals. The consequence of this is that use of electronic instrumentation in soil testing does not necessarily improve the accuracy of the results compared with old-fashioned instruments and may even reduce the accuracy considerably unless the instruments are frequently checked and recalibrated. The moral of all this is that you should always be suspicious of the accuracy of all laboratory tests.

7.11 SUMMARY

1. Laboratory tests are carried out for description and classification of soils, to investigate their basic mechanical properties and to determine values for the stiffness and strength parameters.
2. The principal tests for description and classification are grading by sieving or sedimentation and the Atterberg limit tests which determine the liquid and plastic limits.
3. The principal loading tests are one-dimensional compression (oedometer) tests, shear tests and triaxial tests. These may be drained or undrained and they may be stress controlled or strain controlled.
4. Special loading or unloading stress path tests are carried out in hydraulic triaxial cells. In these tests the axial and radial stresses or strains and the pore pressure can be varied independently to follow the desired stress path.

WORKED EXAMPLES

Example 7.1: Interpretation of a constant head permeameter test   A constant head perme-ameter has a diameter of 100 mm and the standpipe tapping points are 150 mm part. Results of a test on a relatively coarse-grained soil are given in Table 7.1.

Table 7.1

Volume of water collected in 1 min (cm^3)	Difference in standpipe levels (mm)
270	75
220	60
160	45
110	30

The seepage velocity V is given by Eq. (7.5) and the hydraulic gradient i is given by Eq. (7.6). For the first observation,

$$i = \frac{75}{150} = 0.5$$

$$V = \frac{\Delta Q}{A\Delta t} = \frac{270 \times (0.01)^{-3}}{(\pi/4) \times 0.1^2 \times 60} = 5.7 \times 10^{-4} \text{ m/s}$$

Figure 7.11 shows values of V plotted against i. These fall close to a straight line through the origin, which demonstrates that the basic form of Darcy's law (Eq. 7.4) is correct. The coefficient of permeability given by the gradient of the line is

$$k \approx 1 \times 10^{-3} \text{ m/s}$$

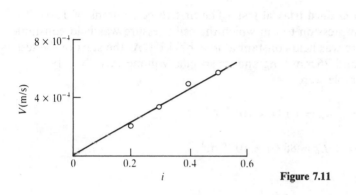

Figure 7.11

Example 7.2: Interpretation of a falling head permeameter test A falling head permeameter has a diameter of 100 mm, the sample is 100 mm long and the area of the standpipe is 70 mm². Results of a test on a relatively fine-grained soil are given in Table 7.2.

At any instant the potential P is the height of water in the standpipe (above the overflow) and $P_0 = 1.60$ mm at $t = t_0$. Figure 7.12 shows the values of $\ln(P_0/P)$ plotted against time. The data points fall close to a straight line. Hence, from Eq. (7.9) the coefficient of permeability is given by

$$k = \frac{aL}{A}\frac{\ln(P_0/P)}{t} = \frac{70 \times (0.001)^2 \times 0.1}{(\pi/4) \times (0.1)^2} \times \frac{0.1}{100} \approx 1 \times 10^{-6} \text{ m/s}$$

Table 7.2

Time (s)	Height of water in standpipe above overflow (m)	$\ln(P_0/P)$
0	1.60	0
60	1.51	0.06
120	1.42	0.12
240	1.26	0.24
480	0.99	0.48

Figure 7.12

Example 7.3: Interpretation of a drained triaxial test   The first three columns of Table 7.3 give data from a drained triaxial compression test in which the cell pressure was held constant at $\sigma_c = 300$ kPa and the pore pressure was held constant at $u = 100$ kPa. At the start of the test the sample was 38 mm in diameter and 76 mm long and its specific volume was $v = 2.19$.

The initial dimensions of the sample were

$$A_0 = \frac{\pi}{4}D_0^2 = 1.134 \times 10^{-3} \text{ m}^2$$

$$V_0 = A_0 L_0 = 88.46 \times 10^{-6} \text{ m}^3$$

At any stage of the test

$$\varepsilon_a = -\frac{\Delta L}{L_0}$$

$$\varepsilon_v = -\frac{\Delta V}{V_0}$$

$$v = v_0(1 - \varepsilon_v)$$

and

$$\sigma_r = \sigma_c = 300 \text{ kPa} \qquad \sigma_a = \sigma_r + \frac{F_a}{A}$$

Table 7.3

Axial force F_a(N)	Change of length ΔL (mm)	Change of volume ΔV (cm³)	ε_s	ε_v	v	q' (kPa)	p' (kPa)
0	0	0	0	0	2.19	0	200
115	−1.95	−0.88	0.022	0.010	2.17	100	233
235	−5.85	−3.72	0.063	0.042	2.10	200	267
325	−11.70	−7.07	0.127	0.080	2.01	264	288
394	−19.11	−8.40	0.220	0.095	1.98	287	296
458	−27.30	−8.40	0.328	0.095	1.98	286	296

where the current area is $A = A_0(1 - \varepsilon_v)/(1 - \varepsilon_a)$. From Eqs (3.5) to (3.8),

$$\varepsilon_s = \varepsilon_a - \tfrac{1}{3}\varepsilon_v$$

and

$$q' = (\sigma'_a - \sigma'_r) = q \qquad p' = \tfrac{1}{3}(\sigma'_a + 2\sigma'_r) = p - u$$

or

$$q' = \frac{F_a}{A} \qquad p' = p_0 + \tfrac{1}{3}q' - u$$

where $p_0 = 300\,\text{kPa}$. The test results are given in the right-hand side of Table 7.3 and are plotted in Fig. 7.13 as O → A.

Figure 7.13

Example 7.4: Interpretation of an undrained triaxial test The first three columns in Table 7.4 give data from an undrained triaxial compression test in which the cell pressure was held

Table 7.4

Axial force F_a (N)	Change of length ΔL (mm)	Pore pressure (kPa)	ε_s	q' (kPa)	p' (kPa)
0	0	100	0	0	200
58	−1.95	165	0.026	50	152
96	−4.29	200	0.056	80	127
124	−9.36	224	0.123	96	108
136	−14.04	232	0.185	98	101
148	−19.50	232	0.257	97	100

constant at $\sigma_c = 300\,\text{kPa}$. At the start of the test the sample was 38 mm diameter and 76 mm long, the pore pressure was $u_0 = 100\,\text{kPa}$ and the specific volume was $v = 2.19$.

For an undrained test $\varepsilon_v = 0$ (by definition), but otherwise the calculations are the same as those given in Example 7.3. The test results are given in the right-hand side of Table 7.4 and are plotted in Fig. 7.13 as O → B.

Example 7.5: Stress paths The left-hand side of Table 7.5 gives the initial states and increments of axial and radial total stresses for a set of drained and undrained triaxial stress path tests. In the drained tests the pore pressure was $u = 0$. The soil can be assumed to be isotropic and elastic so that shearing and volumetric effects are decoupled.

Table 7.5

Sample	σ_a (kPa)	σ_r (kPa)	$d\sigma_a/dt$ (kPa/h)	$d\sigma_r/dt$ (kPa/h)	Drainage	σ_{ae} (kPa)	σ_{re} (kPa)	q'_0 (kPa)	p'_0 (kPa)	q'_e (kPa)	p'_e (kPa)
A	200	200	10	0	Drained	300	200	0	200	100	233
B	200	200	−10	0	Undrained	100	200	0	200	−100	200
C	250	175	−10	−10	Drained	150	75	75	200	75	100
D	250	175	0	−10	Drained	250	75	75	200	175	133

The stress paths corresponding to tests lasting for 10 hours are shown in Fig. 7.14. The right-hand side of Table 7.5 gives the states at the start and at the end of each path. For the undrained test $\delta p' = 0$ (because $\delta \varepsilon_v = 0$ and shear and volumetric effects are decoupled). For the drained tests the changes of q' and p' are found from Eqs (7.13) and (7.14).

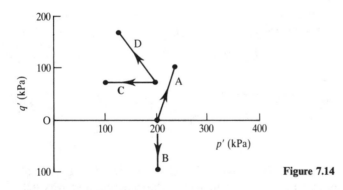

Figure 7.14

FURTHER READING

Atkinson, J. H. and G. Sallfors (1991) 'Experimental determination of soil properties. General Report to Session 1', *Proceedings of 10th ECSMFE, Florence*, Vol. 3, pp. 915–956.

Baldi, G., D. W. Hight and G. E. Thomas (1988) 'A re-evaluation of conventional triaxial test methods', in *Advanced Triaxial Testing of Soil and Rock*, R. T. Donaghe, R. C. Chaney and M. L. Silver (eds), ASTM, STP 977, pp. 219–263.

Bishop, A. W. and D. J. Henkel (1962) *The Triaxial Test*, Edward Arnold, London.

BS 1377 (1991) *Methods of Test for Soils for Civil Engineering Purposes*, British Standards Institution, London.

Head, K. H. (1980) *Manual of Soil Laboratory Testing*, Vol. 1, *Soil Classification and Compaction Tests*, Pentech Press, London.

Head, K. H. (1982) *Manual of Soil Laboratory Testing*, Vol. 2, *Permeability, Shear Strength and Compressibility Tests*, Pentech Press, London.

Head, K. H. (1986) *Manual of Soil Laboratory Testing*, Vol. 3, *Effective Stress Tests*, Pentech Press, London.

EIGHT

COMPRESSION AND SWELLING

8.1 INTRODUCTION

As soils are loaded or unloaded isotropically (i.e. with equal all-round stresses) or anisotropically they will compress and swell. As we saw in Chapter 6, volume changes in soils involve rearrangement of the soil grains and seepage of water as shown in Fig. 8.1.

To account for seepage flow it is necessary to consider the relative rates of loading and drainage as discussed in Sec. 6.9; this is equally true for laboratory tests and for loadings of structures in the ground. In laboratory tests the sample may be loaded undrained and then allowed to consolidate under constant total stress; this is the basis of the conventional incremental loading oedometer test described in Chapter 7. In this case measurements of effective stress can only be made at the end of consolidation when all the excess pore pressures have dissipated (unless the excess pore pressures are measured separately). Alternatively, the loading could be applied at a continuous rate and the excess pore pressures measured. Analysis of continuous loading compression and consolidation tests are discussed in Chapter 14. It is simplest, however, to load samples fully drained at a rate that is slow enough to ensure that any excess pore pressures are negligible so that effective stresses can be determined. I will consider the behaviour of soil during incremental and continuous loading consolidation tests later; for the present I will consider only fully drained states where excess pore pressures are zero. The idealized behaviour described in this chapter is based on experimental data given by Atkinson and Bransby (1978) and by Muir Wood (1991).

(a) Loose soil   (b) Dense soil   **Figure 8.1** Volume changes in soil.

Figure 8.2 Isotropic Compression and swelling.

8.2 ISOTROPIC COMPRESSION AND SWELLING

The general behaviour of soil during isotropic compression and swelling is illustrated in Fig. 8.2. This shows soil in which the grains are loosely packed, initially at p'_0 at O compressed to A, unloaded to B and reloaded through C to D where the grains are more densely packed. This behaviour is similar to that illustrated in Fig. 3.12 and C is a yield point.

Soil compression is primarily caused by rearrangement of the grains and so the stiffness will increase from loose states (where there are plenty of voids for grains to move into) to dense states (where there is much less opportunity for grains to rearrange). As shown in Fig. 8.2, the stress–strain line is curved. Thus the mechanisms of volume change in soils due to rearrangement of the grains accounts for the non-linear bulk stiffness behaviour. For the unloading–reloading loop ABC the soil is very much stiffer (i.e. the volume changes are less) than for first loading because the grains will obviously not 'un-rearrange' themselves on unloading. Behaviour similar to that shown in Fig. 8.2 is also found for soils which have weak grains (such as carbonate or shelly sands) that fracture on loading. In this case most of the compression during first loading is associated with grain fracture but obviously the grains do not 'unfracture' on unloading. From Eq. (3.10) the instantaneous bulk modulus at any point is the gradient of the curve for first loading or for unloading or reloading, given by

$$K' = \frac{dp'}{d\varepsilon_v} \tag{8.1}$$

and the value of K' is not a soil constant.

The behaviour shown in Fig. 8.2 is repeated in Fig. 8.3(a) but plotted as specific volume instead of volumetric strain and with p' plotted horizontally; this is the conventional representation of soil compression and swelling. Figure 8.3(b) shows the same behaviour but now with the stress on a logarithmic scale. In Fig. 8.3(b) the compression and swelling curves from Fig. 8.3(a) are now linear which is a very good approximation for the behaviour of many soils over a wide range of loadings. This idealization is good for most clays and for sands. For coarse-grained soils volume changes during the first loading are often accompanied by fracture of the soil grains and it is usually necessary to apply large stresses (greater than 1000 kPa) to identify the full range of behaviour.

The line OACD corresponding to first loading is known as the normal compression line (NCL) and is given by

$$v = N - \lambda \ln p' \tag{8.2}$$

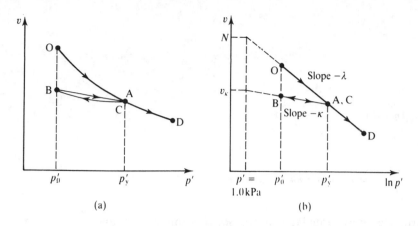

Figure 8.3 Isotropic compression and swelling.

where λ is the gradient and N is the value of v at $p' = 1.0\,\text{kPa}$. The line ABC is known as a swelling line and is given by

$$v = v_\kappa - \kappa \ln p' \qquad (8.3)$$

where κ is the gradient and v_κ is the value of v at $p' = 1.0\,\text{kPa}$. The swelling line ABC meets the normal compression line at C which is a yield point and the yield stress is p_y'. The parameters λ, κ and N are regarded as constants for a particular soil and values for some typical soil types are given in Chapter 9. Soil could be unloaded from any point on the normal compression line and there are any number of swelling lines. For each line there is a particular value of v_κ and a particular value for the yield stress p_y'. Using Eqs (8.2) and (8.3) it is possible to calculate the current specific volume of any isotropically compressed sample given the history of loading and unloading and to calculate the recoverable and irrecoverable volume changes.

From Eq. (8.2), differentiating with respect to p' and dividing by v we have

$$-\frac{dv}{v} = \frac{\lambda}{vp'}\,dp' = d\varepsilon_v \qquad (8.4)$$

and, comparing with Eq. (8.1),

$$K' = \frac{vp'}{\lambda} \qquad (8.5)$$

which is appropriate for first loading. Similarly, for unloading and reloading, we have $K' = vp'/\kappa$. Notice that the bulk modulus K' contains λ or κ which are taken to be constants for a particular soil and vp' which changes during loading and unloading. As a result K' is not a constant and so isotropic compression and swelling lines are non-linear, as shown in Figs 8.2 and 8.3.

8.3 OVERCONSOLIDATION

In Fig. 8.4 the state of a soil during first loading, after deposition, travels down the normal compression line OACD and soil that has been unloaded or reloaded travels on a swelling and recompression line such as ABC characterized by v_κ or p_y'. The state of the soil can reach any point below and to the left of the normal compression line by unloading, but the state cannot reach the region above and to the right. Hence the normal compression line is a boundary to

Figure 8.4 Overconsolidation.

all possible states for isotropic compression: later we will see that this state boundary line forms part of a state boundary surface.

At any state such as B inside the boundary surface the soil is known as overconsolidated and the overconsolidation ratio R_p is given by

$$R_p = \frac{p_y'}{p_0'} \tag{8.6}$$

where p_0' is the current stress and p_y' is a yield point which lies at the intersection of the swelling line through B with the normal compression line. It is very important to notice that I have defined p_y' as the stress at the intersection of the swelling line and the normal compression line and *not* as the stress where the soil was first unloaded; usually, but not always, these will be the same. Notice also that any isotropic state can be described by only *two* of the parameters v, p' and R_p.

For a normally consolidated soil the state lies on the normal compression line and $R_p = 1.0$. Figure 8.5 shows two states, R_1 and R_2, that have the same overconsolidation ratio. From the geometry of the figure, or from Eq. (8.5),

$$\ln R_p = (\ln p_{y1}' - \ln p_{01}') = (\ln p_{y2}' - \ln p_{02}') \tag{8.7}$$

so that the line through R_1 and R_2, where the overconsolidation ratio is the same, is parallel to the normal compression line.

Soils at points N_1 and R_2 have the same current stress, and so would be at the same depth in the ground, but they have very different stiffnesses related to λ and κ respectively. Similarly, soils at points R_2 and N_2 have nearly the same specific volume and water content but, again, they have very different stiffnesses. This means that soil stiffness is not directly related either to the water content or to the current stress (or depth in the ground) and the overconsolidation ratio is an important factor in determining soil behaviour.

Figure 8.5 Overconsolidation ratio.

Figure 8.6 Changes of state due to creep or vibration.

In Fig. 8.6 there are two states at R_1 and R_2 that have the same current stress p'_0 but different overconsolidation ratios because their values of p'_y are different. Normally, for isotropic loading, the state could move from R_1 to R_2 only via the normal compression line through N_1 and N_2: this is equivalent to the yielding and hardening shown in Fig. 3.13 and results in an irrecoverable plastic volume change, δv^p. There are, however, ways in which soils may move from R_1 and R_2 directly. The most important of these are creep in clay soils and vibration or compaction in sands and gravels. Notice that as the state moves from R_1 to R_2 the overconsolidation ratio R_p increases because the value of the yield stress (which is defined as the stress at the intersection of a swelling line with the normal compression line) increases from p'_{y1} to p'_{y2}.

8.4 STATES OF SOILS ON THE WET SIDE AND ON THE DRY SIDE OF CRITICAL

Clays may be normally consolidated or, depending on how far the state is from the normal consolidation line, lightly or heavily overconsolidated, and there is a critical overconsolidation ratio, shown in Fig. 8.7(a), which separates lightly and heavily overconsolidated soils. (We will see later that this critical line corresponds to states at which soil fails during shearing.) The precise value for the critical overconsolidation ratio depends principally on the nature of the soil; most soils will be lightly overconsolidated at $R_p < 2$ and heavily overconsolidated at $R_p > 3$.

Sands and gravels may be loose or dense depending on the position of the state with respect to the critical overconsolidation line, as shown in Fig. 8.7(b). Notice that the state is defined by a combination of specific volume and pressure. In Fig. 8.7(b) the state at A is dense while the state at B is loose although the specific volume at B is smaller than at A: this is because the stress at B is considerably greater than at A. Similarly, in Fig. 8.7(a) the state at A is heavily overconsolidated while the state at B is only lightly overconsolidated although the specific volume at B is smaller than that at A. The regions in which clays are normally consolidated or lightly overconsolidated and sands are loose are said to be on the wet side of the critical line, as shown in Fig. 8.7(c), and the regions where clays are heavily overconsolidated and sands are dense are said to be on the dry side. We will find later that there are fundamental differences in the behaviour of soils when they are sheared from states initially on the wet side or initially on the dry side of the critical line.

Do not misunderstand the terms wet side and dry side. The soil is always either saturated or dry and it is simply that at a given stress, such as p'_c in Fig. 8.7(c), the specific volume (or water content) on the wet side is higher than v_c (i.e. the soil is wetter than at the critical state) while the specific volume (or water content) on the dry side is lower than v_c (i.e. the soil is drier than at the critical state).

Figure 8.7 States of soils on the wet side and on the dry side of critical.

The distance of the initial state from the critical line is a measure of the state of a soil that includes both the current stress and the current volume. This distance may be described either in terms of a stress state parameter S_s or by a volume state parameter S_v. For the state at A in Fig. 8.7(d) these are given by

$$S_s = p'_a/p'_c \qquad \text{or} \qquad \ln S_s = \ln p'_a - \ln p'_c \tag{8.8}$$

$$S_v = v_a - v_c \tag{8.9}$$

Since the critical line and the normal compression line have the same gradient, the state parameters are related by $S_v = \lambda \ln S_s$, and so either may be used to describe the initial state of a soil. If the state is on the critical line, $S_v = \ln S_s = 0$; if the state is on the dry side, S_v and $\ln S_s$ are negative and if the state is on the wet side, S_v and $\ln S_s$ are positive. Notice that the stress state parameter S_s is similar to the reciprocal of the overconsolidation ratio R_p, but it relates the current state to the critical line rather than to the normal compression line. The volume state parameter S_v is similar to the state parameter defined by Been and Jefferies (1985).

8.5 ONE-DIMENSIONAL COMPRESSION AND SWELLING

In the ground the stresses are not generally isotropic as the horizontal and vertical stresses are different. A common case where a relative wide load from an embankment or spread foundation is on a relatively thin layer of clay sandwiched between stiff sand is illustrated in Fig. 8.8. In this case the horizontal strains below most of the embankment are approximately zero, as shown,

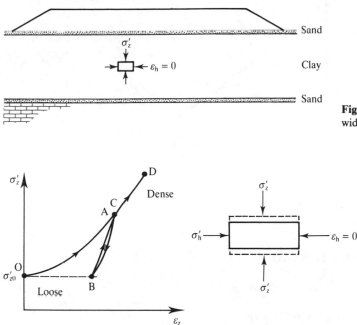

Figure 8.8 One-dimensional states beneath a wide embankment.

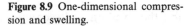

Figure 8.9 One-dimensional compression and swelling.

and the loading is one-dimensional. In the laboratory one-dimensional conditions occur in oedometer tests and in shear box tests before the shear stresses are applied. Although here we are concerned with one-dimensional loading, the conditions below the foundation illustrated in Fig. 8.8 and in the one-dimensional laboratory tests correspond to one-dimensional drainage as well.

The general behaviour of soil during one-dimensional compression and swelling is illustrated in Fig. 8.9. This corresponds to the same sequence of loading, unloading and reloading illustrated in Fig. 8.2, except that the results are shown as vertical stress σ'_z rather than mean stress p' and vertical strain ε_z rather than volumetric strain ε_v; note, however, that for one-dimensional straining where $\varepsilon_h = 0$ we have $\varepsilon_z = \varepsilon_v$. The one-dimensional compression modulus M' is given by

$$M' = \frac{d\sigma'_z}{d\varepsilon_z} \tag{8.10}$$

and, as before, C is a yield point. A parameter often quoted in practice is the one-dimensional coefficient of compressibility m_v given by

$$m_v = \frac{1}{M'} = \frac{d\varepsilon_z}{d\sigma'_z} \tag{8.11}$$

Figure 8.10(a) shows the same behaviour as that in Fig. 8.9 and is equivalent to Fig. 8.3(a) for isotropic compression. Figure 8.10(b) shows the same behaviour with σ'_z plotted to a $\log_{10}$ scale and specific volume replaced by voids ratio. (The axes e and $\log \sigma'_z$ are commonly used in practice for plotting the results of one-dimensional tests and, in Eq. (8.11). $\delta\varepsilon_z = \delta e/(1 + e)$. All the essential features for isotropic compression and swelling described in Sec. 8.2 are repeated for one-dimensional compression and swelling. The principal differences are that the parameter N for isotropic compression is replaced by e_0 and the parameters λ and κ are replaced by C_c and

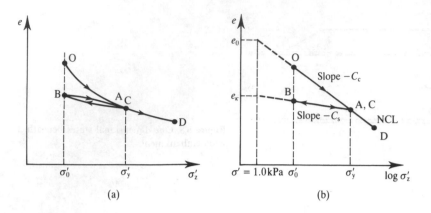

Figure 8.10 One-dimensional compression and swelling.

C_s. The normal compression line OACD is given by

$$e = e_0 - C_c \log \sigma_z'$$ (8.12)

and the swelling and recompression line ABC is given by

$$e = e_\kappa - C_s \log \sigma_z'$$ (8.13)

Since $\delta v = \delta e$ and $\log_{10} x = 0.43 \ln x$ we have $C_c = 2.3\lambda$ and $C_s = 2.3\kappa$.

For overconsolidated soil at a point such as B in Fig. 8.10(a) the overconsolidation ratio R_0 is given by

$$R_0 = \frac{\sigma_y'}{\sigma_0'}$$ (8.14)

where σ_0' is the current stress and σ_y' is the yield point which lies at the intersection of the swelling line through B with the normal compression line. Compare the definition of R_0 for one-dimensional overconsolidation with the definition of R_p in Eq. (8.6) for isotropic over-consolidation.

During the increase and decrease of σ_z' in one-dimensional loading and unloading the horizontal stresses σ_h' change since ε_h is held constant and the variations of σ_z' and σ_h' are illustrated in Fig. 11(a). The ratio

$$K_0 = \frac{\sigma_h'}{\sigma_z'}$$ (8.15)

is known as the coeffcient of earth pressure at rest (i.e. corresponding to zero horizontal strain) and the variation of K_0 with overconsolidation ratio R_0 is illustrated in Fig. 8.11(b). For states OACD on the normal compression line $R_0 = 1$ and the value of K_0 is K_{0nc} for normally consolidated soil: for many soils this can be approximated by

$$K_{0nc} = 1 - \sin \phi_c'$$ (8.16)

where ϕ_c' is the critical friction angle (see Chapter 9). For overconsolidation states ABC the value of K_0 increases with overconsolidation and K_0 may well exceed 1.0 as the horizontal stress

Figure 8.11 Horizontal stresses during one-dimensional loading and unloading.

exceeds the vertical stress at large values of R_0. Figures 8.11(a) and (b) illustrate substantial hysteresis in K_0 during unloading and reloading, but if this is neglected then K_0 is found to vary with R_0 and an approximate empirical relationship is

$$K_0 = K_{0nc}\sqrt{R_0} \tag{8.17}$$

During one-dimensional loading and unloading σ'_z and σ'_h are generally unequal and so there are shear stresses in the soil and any comparison between isotropic and one-dimensional compression and swelling will have to take account of the shear stresses. The link between these can be developed by going back to Sec. 3.2, noting also that one-dimensional loading is a special case of plane strain. From Eqs (3.5) and (3.6) with $\sigma'_a = \sigma'_z$ and $\sigma'_r = \sigma'_h$, and making use of Eq. (8.14), we have

$$q' = \sigma'_z(1 - K_0) \tag{8.18}$$

$$p' = \tfrac{1}{3}\sigma'_z(1 + 2K_0) \tag{8.19}$$

Figure 8.12 shows the behaviour of soil in isotropic and one-dimensional compression and swelling together; the subscripts 1 refer to one-dimensional behaviour. These show normal compression lines OACD and $O_1A_1C_1D_1$ with the same gradients $-\lambda$ and values of v at $p' = 1$ kPa of N and N_0. The swelling and recompression lines ABC and $A_1B_1C_1$ have approximately the same gradients, $-\kappa$, and the same yield stresses, p'_y, but different values of v_κ. (The gradients κ are actually slightly different because the value of K_0 changes during one-dimensional swelling and recompression.)

8.6 LABORATORY DEMONSTRATIONS OF COMPRESSION AND SWELLING OF SOILS

The most convenient apparatus to demonstrate compression and swelling is the simple oedometer using weights to apply normal stresses and a dial gauge to measure strains. It is not so easy, however, to find a suitable soil as sands do not compress very much, while clays have low permeability and take a long time to compress. A convenient material is a dry sand with weak and friable grains, such as carbonate (shelly) sand. Alternatively, kaolin clay consolidates reasonably quickly (in a standard oedometer consolidation of kaolin takes about 15 min for each stage) and it is reasonably compressible.

Figure 8.12 Soil behaviour during isotropic and one-dimensional compression and swelling.

8.7 SUMMARY

1. The basic mechanism of compression in soils is by rearrangement of the grains. In coarse-grained soils this may be accompanied by fracturing of the soil grains and in fine-grained soils by compression or swelling of clay particles.
2. The behaviour of soil during isotropic compression and swelling is given by

$$v = N - \lambda \ln p' \tag{8.2}$$

$$v = v_\kappa - \kappa \ln p' \tag{8.3}$$

The parameters λ, κ and N are constants for a particular soil. Equation (8.2) is for normally consolidated soil and Eq. (8.3) is for overconsolidated soil.
3. Equations (8.2) and (8.3) demonstrate that the stiffness of soil is non-linear (i.e. the bulk modulus is not a constant) when it is both normally consolidated and overconsolidated.
4. Equation (8.2) represents the normal compression line. The state of a soil cannot usually lie outside this line and moves below the line on unloading when the soil becomes overconsolidated. The overconsolidation ratio R_p is given by

$$R_p = \frac{p'_y}{p'_0} \tag{8.6}$$

where p'_y is the current yield stress.
5. Normally the state of soil is changed only by loading and unloading and the states moves on the current swelling and recompression line or on the normal compression line. The

state of a clay may also change due to creep and the state of a sand may change due to vibration or compaction.

6. There is a critical overconsolidation ratio line which separates the wet side from the dry side. Lightly overconsolidated clays and loose sands are on the wet side of the critical line while heavily overconsolidated clays and dense sands are on the dry side.

7. The initial state can be described by a stress state parameter S_s or a volume state parameter S_v which give the distance of the initial state from the critical overconsolidation line.

8. The behaviour of soil during one-dimensional compression and swelling is similar to that for isotropic loading and is given by

$$e = e_0 - C_c \log \sigma'_z \tag{8.12}$$

$$e = e_\kappa - C_s \log \sigma'_z \tag{8.13}$$

The parameters e_0, C_c and C_s are constants for a particular soil.

WORKED EXAMPLES

Example 8.1: Analysis of an isotropic compression test   Table 8.1 gives results obtained from an isotropic test. The data are shown plotted in Fig. 8.13. Scaling from the diagram, $\lambda = 0.20$ and $\kappa = 0.05$. Projecting the lines back to $p' = 1.0 \, \text{kPa}$ (i.e. $\ln p' = 0$), $N_0 = 3.25$ and $v_\kappa = 2.22$. The bulk modulus K' is not a constant: from Eq. (8.1), for the second and last increments between $p' = 60 \, \text{kPa}$ and $p' = 200 \, \text{kPa}$,

$$K' = \frac{\Delta p'}{\Delta \varepsilon_v} = \frac{200 - 60}{-(2.19 - 2.43)/2.43} = 1.42 \, \text{Mpa}$$

$$K' = \frac{\Delta p'}{\Delta \varepsilon_v} = \frac{60 - 200}{-(2.01 - 1.95)/1.95} = 4.55 \, \text{MPa}$$

Table 8.1

Mean effective stress p' (kPa)	ln p' (kPa)	Specific volume v
20	3.00	2.65
60	4.09	2.43
200	5.30	2.19
1000	6.91	1.87
200	5.30	1.95
60	4.09	2.01

Figure 8.13

Example 8.2: Determination of soil behaviour during isotropic compression A soil has the parameters $\lambda = 0.20$, $\kappa = 0.05$ and $N = 3.25$. A sample is subjected to the sequence of isotropic loading and unloading given in the second column in Table 8.2. At each stage the overconsolidation ratio R_p is given by Eq. (8.6). For the normally consolidated state ($R_p = 1$) the specific volume is given by Eq. (8.2). At the point C, the specific volume is given by both Eqs (8.2) and (8.3) and hence

$$v_\kappa = N - (\lambda - \kappa) \ln p' = 3.25 - (0.20 - 0.05) \ln 600 = 2.29$$

For the overconsolidated states the specific volume is given by Eq. (8.3).

Table 8.2

Point	Mean effective stress p' (kPa)	Overconsolidation ratio R_p	Specific volume v
A	60	1	2.43
B	200	1	2.19
C	600	1	1.97
D	300	2	2.01
E	150	4	2.04

REFERENCE

Been, K. and M. G. Jefferies (1985) 'A state parameter for sands', *Geotechnique*, **35**, 2, 99–112.

FURTHER READING

Atkinson, J. H. and P. L. Bransby (1978) *The Mechanics of Soils*, McGraw-Hill, London.
Muir Wood, D. M. (1991) *Soil Behaviour and Critical State Soil Mechanics*, Cambridge University Press, Cambridge.

CHAPTER NINE

CRITICAL STATE STRENGTH OF SOIL

9.1 BEHAVIOUR OF SOIL IN SHEAR TESTS

In simple terms the strength of a material is the maximum shear stress that it can sustain; materials loaded just beyond the maximum stress will fail. Failure may be sudden and catastrophic leading to a complete loss of strength (which is what happens when you break a piece of blackboard chalk) or it may lead to a very large plastic straining (which is what happens if you mould plasticine). For most soils failure of slopes and foundations involves large plastic straining without complete loss of strength and failing soil structures can usually be stabilized by unloading them.

The essential features of soil strength can most easily be seen in ideal shearing tests, as illustrated in Fig. 9.1. The shear and normal effective stresses are τ' and σ' and, at a particular stage of the test, there are increments of strain $\delta\gamma$ and $\delta\varepsilon_v$. These are similar to the conditions in the direct shear box test and the simple shear test described in Chapter 7 and in soil in thin slip surfaces that occur during failure of slopes and foundations, as described in Chapters 18 and 19. The conventional direct and simple shear tests are, however, not ideal because the stresses and deformations are likely to be non-uniform and the states of stress and strain are not completely defined by the measurements on only one plane. Although a shear test is not ideal for measuring soil properties it is, however, convenient for demonstrating the basic characteristics of soil strength.

Typical stress–strain curves for soils on the wet side of critical (i.e. normally consolidated or lightly overconsolidated clays or loose sands marked W) and for soils on the dry side (i.e. heavily overconsolidated clays or dense sands marked D), tested drained with constant σ', are shown in Fig. 9.1(b) and the corresponding volumetric strains are shown in Fig. 9.1(c). (Remember the distinctions between the wet side of the critical overconsolidation line and the dry side, discussed in Sec. 8.4.) The behaviour shown in Fig. 9.1 is typical for normally consolidated or overconsolidated clays as well as for loose or dense sands. Soils on the wet side compress as the shear stresses increase while soils on the dry side dilate (expand) after a small compression. Both ultimately reach states at which the shear stress is constant and there are no more volumetric strains. Soils on the dry side reach peak shear stresses before reaching the ultimate

Figure 9.1 Typical behaviour of soils in drained shear tests.

state. At any stage of shearing the angle of dilation ψ (see Sec. 2.6) is defined by

$$\tan \psi = -\frac{d\varepsilon_v}{d\gamma} \tag{9.1}$$

This is the gradient of the volume change curve as shown in Fig. 9.1(c) and it also gives the direction of movement of the top of the sample as shown in Fig. 9.1(a). The negative sign is introduced into Eq. (9.1) so that dilation (negative volumetric straining) is associated with positive angles of dilation.

Figure 9.1(d) shows the change of voids ratio e rather than the volumetric strains shown in Fig. 9.1(c), although, of course, they are related. Both samples have the same effective normal stresses but the initial voids ratio of the sample on the wet side is higher than that of the sample

Figure 9.2 Compression and dilation during shearing.

on the dry side. Notice, however, that both samples reach their ultimate states at the same voids ratio e_f.

As volume changes in soils are principally due to rearrangement of particles it is easy to see why soils on the wet side compress while soils on the dry side dilate. In Fig. 9.2 the grains of the loose or normally consolidated soil at W are spaced well apart and, on shearing, they can move into the neighbouring void spaces, while the grains of the dense or overconsolidated soil at D must move apart during shear. This is an example of the coupling between shear and volumetric effects in soils.

9.2 PEAK, ULTIMATE AND RESIDUAL STATES

As shown in Fig. 9.1, soils initially on the dry side of the critical line reach peak shear stress states before the ultimate state. The peak state will normally be reached at strains of the order of 1 per cent while the ultimate state will be reached after strains greater than 10 per cent (in some soils the ultimate states are not reached until the strains have exceeded 50 per cent or so). Notice that the peak state coincides with the point of maximum rate of dilation (i.e. at maximum ψ). Soils on the wet side compress throughout, shearing up to the ultimate state, and there is no peak.

For soils that have a peak shear stress it is not easy to decide whether the strength of the soil—the maximum shear stress it can sustain—should be the peak stress that can be sustained only for relatively small strains or the ultimate state. I will leave this question for the time being and, for the present, I will discuss the conditions at the peak state and the conditions at the ultimate state separately.

There is another aspect of soil shearing that must be considered here and that is the development of residual strength at very large displacements on slip planes (Skempton, 1964). Figure 9.3 illustrates the behaviour of a sand and a plastic clay soil over large displacements; note the logarithmic scale, which allows the diagram to represent displacements exceeding 1 m. (Tests of this kind can be carried out in a direct shear box by moving the box backwards and forwards or in a special ring shear apparatus in which an annulus of soil can be sheared continuously.) The behaviour illustrated is for tests in which the effective stresses and the initial states were chosen so that the peak and ultimate states of the clay and the sand soil happened to be the same. At the ultimate state, at displacements of about 10 mm corresponding to shear strains of about 10 per cent as shown in Fig. 9.1, the movements of grains are essentially turbulent, involving relative movements and rotations of both clay and sand grains. At larger displacements, however, the strains become localized into distinct zones of intense shearing and the shear stresses applied to the clay soil decrease.

Figure 9.3 Residual strength of clay at very large displacements.

The lowest shear stress reached after very large displacements is called the residual state. It is associated with laminar flow of flat clay grains which have become orientated parallel to the rupture zone, as illustrated in Fig. 9.3. In sands and other soils with rotund (i.e. not flat) grains, there is no opportunity for laminar flow and the residual state is the same as the ultimate state. In clays, the residual state may be as little as 50 per cent of the ultimate state and it is important for the design of works on old landslides and for determining side friction on driven piles.

9.3 CRITICAL STATES

We now come to the essence of soil mechanics, which is the ultimate or critical state. The idealized behaviour described in this chapter is based on experimental data given by Atkinson and Bransby (1978) and by Muir Wood (1991). From Figs 9.1 and 9.3 the critical state is the state reached after strains of at least 10 per cent and is associated with turbulent flow. The relationships between the shear stress, the normal stress and the voids ratio of soils at the critical states are illustrated in Fig. 9.4.

Figure 9.4(a) and (b) shows the critical state line (CSL). This shows that, at the critical state, there is a unique relationship between the shear stress, the normal stress and the voids ratio. Figure 9.4(c) is the same as Fig. 9.4(b) but with the normal stress on a logarithmic scale. Also shown on Fig. 9.4(c) is the one-dimensional normal compression line from Fig. 8.10(b).

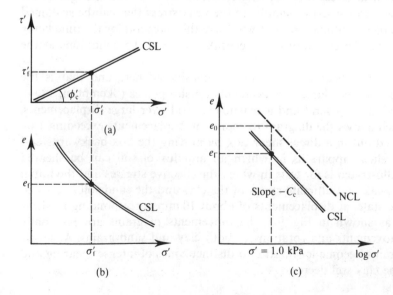

Figure 9.4 Critical states of soils.

The critical state line is given by

$$\tau_f' = \sigma_f' \tan \phi_c' \tag{9.2}$$

$$e_f = e_\Gamma - C_c \log \sigma_f' \tag{9.3}$$

where the subscripts f denote that the stresses and the voids ratio are those at ultimate failure at the critical state. In Fig. 9.4(c) the normal compression and critical state lines are parallel and both have the same gradient, C_c. The parameter e_Γ defines the position of the critical state line in the same way that e_0 defines the position of the normal compression line. Equation (9.2) is the Mohr–Coulomb failure criterion discussed in Sec. 3.3 with zero cohesion ($c' = 0$) and ϕ_c' is the critical friction angle. The critical state line shown in Fig. 9.4(c) is directly above the critical overconsolidation line shown in Fig. 8.7. (The height of the critical state line above the critical overconsolidation line is τ_f' given by Eq. (9.2).) Later, in Chapter 11, we will see how the state of a soil initially on the wet side or the dry side moves towards the critical state line during shearing.

It is essential to emphasize that at the critical state soil continues to distort (i.e. suffer shear strains) without any change of shear stress or normal stress or voids ratio (i.e. it is distorting at constant state) and the strains are associated with turbulent flow. The essential features of the critical states are that, during shearing, all soils will ultimately reach their critical states (provided that the flow remains turbulent) and the ultimate or critical states are independent of the initial states. Thus, in Fig. 9.1, the ultimate or critical shear stresses τ_f' are the same for the soils initially on either the wet or the dry sides of critical, because they have the same normal effective stress σ_f' and the voids ratios e_f at the critical states will also be the same. Later we will see how we can explain fully the behaviour of soils from knowledge of their initial and ultimate states.

The existence of unique critical states for soils is, at first sight, surprising, but it is quite logical. Firstly, during continuous shear straining any soil must ultimately reach a constant state because, if it did not, it would continue to dilate or compress and strengthen or weaken indefinitely, which is, of course, nonsense. During shearing from the initial to the critical states there will be relatively large strains and the soil will be essentially reworked or reconstituted by the shear straining. Thus the soil will forget its initial state and it is reasonable to suppose that the new, reconstituted, soil will achieve unique states independent of the initial states.

The general critical state lines illustrated in Fig. 9.4 are a very good idealization for the ultimate or critical states of most clays and sands. For coarse-grained soils volume changes during first loading and during shearing are often accompanied by fracture of the soil grains, and it is often necessary to apply large stresses (greater than 1000 kPa) to identify the full range of behaviour.

9.4 UNDRAINED STRENGTH

The ultimate or critical state strength of soil given by Eq. (9.2) relates the ultimate shearing resistance to the corresponding normal effective stress. This can be used to determine soil strength provided that the pore pressure is known so that σ' ($= \sigma - u$) can be calculated. Pore pressures in the ground will generally only be determinable for cases of drained loading and the strength for undrained loading—the undrained strength—must be calculated differently.

Figure 9.5(a) and (b) shows the critical state line for soil and is the same as Fig. 9.4(a) and (b). Figure 9.5(c) combines these and shows the corresponding relationship between the critical state shear stress and the voids ratio: this shows the strength decreasing with increasing voids ratio. Since the voids ratio is simply related to water content in saturated soil (see Eq. 5.6) Fig.

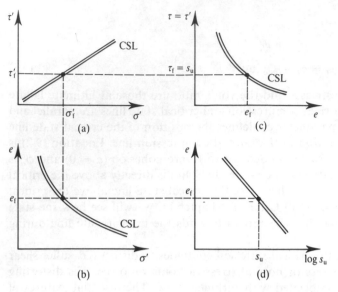

Figure 9.5 Undrained strength of soil.

9.5(d) shows that soil strength decreases with increasing water content and it is common knowledge that soils become softer and weaker as they become wetter. Notice that the soil is either saturated, or dry, so the change of strength with changing water content (i.e. with wetting or drying) is nothing to do with lubrication: it is a result of the change in effective stress which is related to the change of voids ratio.

Figure 9.5(c) shows that there is a unique relationship between the critical shear stress (i.e. the ultimate strength) and the voids ratio or water content. This means that for any undrained loading—loading at constant voids ratio—the strength is independent of any changes in the total normal stress: this is called the undrained shear strength s_u. For undrained loading the undrained strength is

$$\tau_f = s_u \tag{9.4}$$

which is the Tresca failure criterion discussed in Sec. 3.3. From Eqs (9.2) and (9.3) and noting that $\tau_f' = \tau_f = s_u$, we have

$$\log\left(\frac{s_u}{\tan \phi_c'}\right) = \frac{e_\Gamma - e}{C_c} \tag{9.5}$$

and the relationship between undrained strength s_u and voids ratio is illustrated in Fig. 9.5(d).

In dealing with the undrained shear strength s_u the important thing to remember is that it applies only for loading or unloading at constant volume (i.e. undrained). As a result, s_u depends only on the voids ratio or water content and is independent of the total normal stress. This means that we can measure s_u in any field or laboratory test, with any total stresses, provided that the soil remains undrained.

For undrained loading the pore pressures are not easily determined and so routine design calculations are done using total, not effective, stresses. This does not matter because once the undrained strength s_u has been determined it does not change so long as the voids ratio does not change. For this reason, analyses for undrained loading using the undrained strength are called total stress analyses. For drained loading, however, the pore pressures and the effective stresses are determinable and calculations that are carried out using the critical friction angle ϕ_c' are called effective stress analyses.

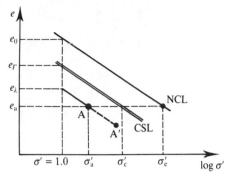

Figure 9.6 Parameters for normalizing shear test results.

9.5 NORMALIZING

Representation of the critical state line, as in Figs 9.4 and 9.5, is relatively straightforward because, at the critical state τ_f', σ_f' and e_f are uniquely related and there is only one critical state line. When we come to deal with peak states and other states before the critical, the situation is a little more complex and it will be convenient to have a method of normalizing stresses and voids ratios or specific volumes to simplify the presentation.

In Fig. 9.6 there is a point A where the state is σ_a' and e_a and there may also be some shear stresses (not necessarily at the critical state) τ_a'. In Sec. 8.3 we found that the overconsolidation ratio or the current state was an important factor in determining soil behaviour and so all the states with the same overconsolidation ratio or the same state parameters should ideally have the same equivalent state after normalization. This can be achieved in a variety of ways and the two most common are illustrated in Fig. 9.6.

We have already seen that the positions of the normal compression and critical state lines are defined by the parameters e_0 and e_Γ and so the line of constant overconsolidation ratio containing A and A′ is given by

$$e_\lambda = e_a + C_c \log \sigma_a' \tag{9.6}$$

Notice that e_λ contains both e_a and σ_a' and e_λ decreases with increasing overconsolidation ratio.

Figure 9.7(b) shows the one-dimensional normal compression and critical state lines plotted with axes e_λ and τ' normalized with respect to the current stress σ'. Both lines appear as single points; at the normal compression point $\tau'/\sigma' = 0$ and $e_\lambda = e_0$ while at the critical state point $\tau'/\sigma' = \tan \phi_c'$ and $e_\lambda = e_\Gamma$. It seems fairly obvious that there will be important states between these, represented by the broken line, and we will explore these later.

A second method of normalizing is to make use of an equivalent stress. In Fig. 9.6 the stress σ_c' is on the critical state line at the same voids ratio as A and we have

$$\log \sigma_c' = \frac{e_\Gamma - e_a}{C_c} \tag{9.7}$$

(There is another equivalent pressure σ_e' on the normal compression line which is often used as a normalizing parameter. In this book I want to use σ_c' because the critical state line is unique for a given soil, while there are different normal compression lines for isotropic and one-dimensional compression and the position of the normal compression lines of natural soils can be influenced by cementing, structure and other effects.) Figure 9.7(a) shows the normal compression and critical state lines plotted with normalized stresses τ'/σ_c' and σ'/σ_c'. Again both lines appear as single points and the broken line corresponds to the broken line in Fig. 9.7(b). The position of the critical state line is determined by $\tau'/\sigma' = \tan \phi_c'$ and $\sigma'/\sigma_c' = 1.0$. From the

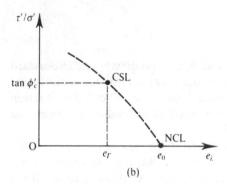

Figure 9.7 Normalized critical state and normal consolidation lines.

geometry of Fig. 9.6 the position of the normal compression line is given by

$$\log\left(\frac{\sigma'_e}{\sigma'_c}\right) = \frac{e_0 - e_\Gamma}{C_c} \tag{9.8}$$

9.6 CRITICAL STATE STRENGTH OF SOILS MEASURED IN TRIAXIAL TESTS

So far we have considered strength of soils in ideal shear tests. As it is impossible to control and measure pore pressures in the conventional shear box apparatus the tests were drained so the pore pressure was zero and total and effective stresses were equal. We also considered undrained tests in which pore pressures were not measured and the undrained strength was related to the constant voids ratio. A more common and more useful test to examine soil behaviour is the triaxial test described in Chapter 7. In the triaxial test a cylindrical sample is subjected to total axial and radial stresses while the pore pressures and the sample volume can be controlled and measured independently so that it is possible to determine the effective stresses and the strains.

Relationships between stresses and strains in shear and triaxial tests were discussed in Chapter 3. For shear tests the shear and normal stresses and strains are τ', σ', γ and ε_v and for triaxial tests the equivalent parameters are q', p', ε_s and ε_v; these can be related through Mohr circle constructions, as described in Chapter 3.

All the features of soil behaviour in shear tests shown in Fig. 9.1 are seen in the results of triaxial tests plotted as q' against ε_s and ε_v against ε_s. In triaxial tests soils reach ultimate or critical states where they continue to distort at a constant state (i.e. with constant effective stresses and constant volume) and soils initially on the dry side of the critical state line have peak states before the critical state is reached.

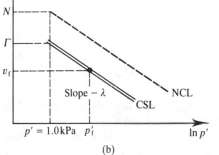

Figure 9.8 Critical state line for triaxial tests.

The critical state lines obtained from drained and undrained triaxial tests are shown in Fig. 9.8, which may be compared with Fig. 9.4 showing the critical state lines for shear tests. The critical lines in Fig. 9.8 are given by

$$q'_f = Mp'_f \tag{9.9}$$

$$v_f = \Gamma - \lambda \ln p'_f \tag{9.10}$$

where, as before, the subscripts f denote ultimate failure at the critical states. Comparing Eq. (9.9) with Eq. (9.2), the critical stress ratio M is equivalent to the critical friction angle ϕ'_c. In Fig. 9.8(b) the gradients of the critical state line and the isotropic normal compression line are λ and the lines are parallel and the gradient of the critical state line is the same for triaxial compression and extension. For the parameters M and Γ, however, it is necessary to use

Table 9.1

Soil	LL	PL	Typical soil parameters					
			λ	Γ	N	M	ϕ'	κ/λ
Fine-grained clay soils								
London clay	75	30	0.16	2.45	2.68	0.89	23°	0.39
Kaolin clay	65	35	0.19	3.14	3.26	1.00	25°	0.26
Glacial till	35	17	0.09	1.81	1.98	1.18	29°	0.16
Coarse-grained soils								
River sand			0.16	2.99	3.17	1.28	32°	0.09
Decomposed granite			0.09	2.04	2.17	1.59	39°	0.06
Carbonate sand			0.34	4.35	4.80	1.65	40°	0.01

Figure 9.9 Parameters for normalizing triaxial test results.

subscripts c and e to distinguish between critical states in compression and extension, and for most soils the values of both Γ_c and Γ_e and M_c and M_e differ. The parameters λ, Γ and M (or ϕ') for triaxial compression are regarded as constants for a particular soil and values for some typical soils are given in Table 9.1.

Results of triaxial tests may be normalized like the results of shear tests. The normalizing parameters, shown in Fig. 9.9, are the critical pressure p'_c and the equivalent specific volume v_λ; these are comparable to σ'_c and e_λ in Fig. 9.6. (The equivalent pressure on the isotropic normal compression line p'_e is often used as a normalizing parameter for triaxial tests but, again, I want to use p'_c because the critical state line is unique.) From the geometry of Fig. 9.9,

$$v_\lambda = v_a + \lambda \ln p'_a \tag{9.11}$$

$$\ln p'_c = \frac{\Gamma - v_a}{\lambda} \tag{9.12}$$

(a)

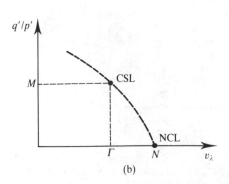

(b)

Figure 9.10 Normalized critical state and normal consolidation lines.

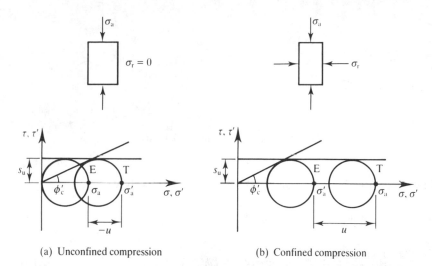

(a) Unconfined compression   (b) Confined compression

Figure 9.11 Undrained strength in compression tests

Figure 9.10 shows critical state and normal compression lines normalized with respect to p'_c and v_λ: these correspond to Fig. 9.7 for shear tests. Again a broken line has been drawn representing important states between the normal compression and critical state lines; we will consider these states in later chapters. Note that, for triaxial tests, there will be two critical state lines, one for compression and one for extension.

The undrained strength s_u is uniquely related to the voids ratio, and hence to the specific volume. From Eqs (9.9) and (9.10), noting that $s_u = \frac{1}{2}(\sigma'_a - \sigma'_r)_f = \frac{1}{2}q'_f$ we have

$$\ln\left(\frac{2s_u}{M}\right) = \frac{\Gamma - v}{\lambda} \tag{9.13}$$

which is comparable to Eq. (9.5). Undrained strength may be measured in unconfined compression tests (i.e. tests with $\sigma_r = 0$) or in triaxial tests with any confining pressure provided that the voids ratio does not change. Figure 9.11 shows Mohr circles of total and effective stress for confined and unconfined compression tests on samples with the same voids ratio. The Mohr circles of effective stress are identical; they both touch the lines given by $\tau_f = s_u$ and $\tau'_f = \sigma'_f \tan \phi'_c$. The Mohr circles of total stress have the same diameter (because the voids ratios of the samples are the same) but they are in different positions, so the pore pressure in the unconfined compression test sample is negative. It is this negative pore pressure that produces positive effective stresses and gives rise to the unconfined compressive strength; this accounts for the strength of a sandcastle and the stability of a trench with steep sides.

9.7 RELATIONSHIPS BETWEEN STRENGTH MEASURED IN SHEAR AND TRIAXIAL TESTS

The relationships between stress ratios in shear and triaxial tests using the Mohr circle constructions were introduced in Chapter 2 and these can be used to relate the results of triaxial

Figure 9.12 Stress ratios in triaxial tests.

and shear tests. From Fig. 9.12 with $t' = \frac{1}{2}(\sigma'_a - \sigma'_r)$ and $s' = \frac{1}{2}(\sigma'_a + \sigma'_r)$ we have

$$\frac{t'}{s'} = \sin \phi'_{mob} = \frac{(\sigma'_a - \sigma'_r)}{(\sigma'_a + \sigma'_r)} \tag{9.14}$$

$$\frac{\sigma'_a}{\sigma'_r} = \frac{(1 + \sin \phi'_{mob})}{(1 - \sin \phi'_{mob})} = \tan^2(45 + \tfrac{1}{2}\phi'_{mob}) \tag{9.15}$$

and, at the critical state $\phi'_{mob} = \phi'_c$. Relationships between ϕ'_c and M can be obtained from Eqs (9.9) and (9.15) with $q' = \sigma'_a - \sigma'_r$ and $p' = \frac{1}{3}(\sigma'_a + 2\sigma'_r)$, noting that for compression $\sigma'_a > \sigma'_r$ while for extension $\sigma'_a < \sigma'_r$, so that in Eq. (9.15) σ'_a/σ'_r for compression must be replaced with σ'_r/σ'_a for extension. Readers are invited to work through the algebra and demonstrate that

$$M_c = \frac{6 \sin \phi'_c}{3 - \sin \phi'_c} \tag{9.16}$$

$$M_e = \frac{6 \sin \phi'_c}{3 + \sin \phi'_c} \tag{9.17}$$

The critical friction angle ϕ'_c is approximately the same for triaxial compression and extension, so Eqs (9.16) and (9.17) demonstrate that M_c and M_e are not equal and $M_c > M_e$.

9.8 SIMPLE EXPERIMENTAL INVESTIGATIONS OF CRITICAL STATES

In any theoretical or experimental study of soil, and in many design studies, it is essential to determine the position of the critical state line as accurately as possible. This is needed to determine the ultimate strength for many of the design analyses described in Chapters 18 to 23 and it is also required to determine the ultimate states of soil samples. It is important to be able to distinguish between states on the wet side of critical from states on the dry side of critical, and the critical state parameters λ and Γ (or C_c and e_r) are required for normalizing soil test data.

However, if you try to measure critical states of soils in the conventional shear or triaxial tests described in Chapter 7 or if you use test results obtained by other people, you must be very careful to ensure that the samples really have reached their critical states, as defined in Sec. 9.3. Remember that for soils to reach their critical states they must be straining with no change of state (i.e. at constant shear stress, constant effective normal stress and constant volume) and with turbulent flow. This means that, if the stresses or pore pressures change at all or if there are any volume changes, the states measured in the tests will not be the critical states. Very often soil tests are terminated when the apparatus runs out of travel, usually at strains of 20 per cent or so. In many cases these strains are not large enough to reach the critical states in soils initially on the wet side of critical and are sufficient to cause slip planes to form in soils initially on the dry side. As discussed later, if there are any distinct slip surfaces in a test sample

Figure 9.13 Rotating cylinder tests for ϕ'_c.

Valve

Figure 9.14 Submerged cone test.

stresses and strains become non-uniform and cannot be measured reliably. We will discover later (in Chapter 11) that it is possible to find the critical states of soils from tests on normally consolidated and overconsolidated samples by considering the stresses and volume changes at strains before the critical state is reached.

Possibly the best tests to determine the critical states of soils should be carried out on lightly overconsolidated samples for which the initial specific volume or voids ratio is close to the critical state value. Some people will tell you that soils do not reach unique critical states or that the critical state lines are curved, but usually their test data are suspect because the samples were not at their rigorously defined critical states.

There are some simple experiments that can be done to illustrate the critical states of soils and to obtain reasonable values of soil parameters. Because these simple experiments do not control pore pressures or drainage, tests on sands will be drained and will examine the critical friction angle ϕ'_c, while tests on clays will be undrained and will examine the undrained strength s_u.

We will see later that if there is no seepage the critical angle i_c of a failing slope is equal to the critical friction angle ϕ'_c, and so observation of slopes is a convenient method of determining the friction angle of soil. One test is to put dry sand in a horizontal rotating cylinder; as the cylinder rotates the angle of the continuously failing slope is the critical angle, as shown in Fig. 9.13. Another test is to pour dry sand into a cone and measure the cone angle. A better test is to do this under water, as shown in Fig. 9.14 (you must pour saturated sand through water in this test), and the slope angle is the same as for dry sand. (If you open the valve and drain water from the container the slope will slump to a shallower angle as water drains from the sand cone.)

The undrained strength of normally consolidated or lightly overconsolidated clays can be measured in unconfined compression tests, but it is time consuming to prepare saturated samples at different water contents. Try to obtain samples of the same clay at different water contents and relate the undrained strength to the water content and the voids ratio. While doing these tests examine the strength by squeezing the soil between your fingers and pressing in your fingernail. It is very difficult to measure the undrained critical state strength of heavily overconsolidated clays because they usually fail in a brittle manner with distinct slip planes.

(a) $c' > 0$ (b) $c' = 0$ (c) $c' < 0$

Figure 9.15 Assessment of true cohesion in soils.

9.9 TRUE COHESION IN SOILS

In Figs 9.4(a) and 9.8(a) the critical state lines were drawn passing through the origin, so that the cohesion c' is zero and, at the critical state, soil has no strength when the effective normal stress is zero. If soils are cemented so the grains are glued together they will have a cohesion, but the strains required to reach the critical state are enough to break these cemented bonds.

Critical state lines obtained from the results of laboratory tests on soils always pass through the origin, at least within the accuracy of the results, which is typically about ± 2 to 3 kPa. It is very difficult to measure directly soil strength at zero effective stress. Some materials, such as dry sand, sugar and grain, are obviously cohesionless and you can pour them like water (although they will form cones), but it is not so obvious that fine-grained materials, such as clays, dry cement and flour, are cohesionless. The problem is that any moisture present will give rise to pore suctions which will raise the effective stresses, and hence the strength.

You can only really examine true cohesion in soils if the pore pressures are zero, which is clearly the case in dry materials. Dry flour has no cohesion if it is loose, because you can blow it away, but if you compact it by squeezing it in your hand it has a small strength. This is a result of the relatively large specific surface of finely ground flour.

The pore pressures in saturated fine-grained soils become zero if a sample is submerged in water. Figure 9.15 illustrates the behaviour of initially cylindrical samples of soil with different cohesions after they have been submerged in water. (The samples should be completely reconstituted so that any cementing is destroyed.) If the cohesion is zero as in Fig. 9.15(b) the sample forms a cone. If the cohesion is positive as a result of small interparticle attractions the sample will remain as a cylinder, as shown in Fig. 9.15(a). If, however, the water becomes dirty, this must mean that there were small interparticle repulsive forces and the true cohesion was negative. Each of the three characteristic types of behaviour shown in Fig. 9.15 are observed in tests on soils (Atkinson, Charles and Mhach, 1990). Even though the true cohesion in soils may be positive or negative the values are usually very small, only a few kiloPascals, which is too small to measure reliably in conventional laboratory tests.

9.10 ESTIMATION OF THE CRITICAL STATE STRENGTH PARAMETERS FROM CLASSIFICATION TESTS

At the critical state, after large strains, soil is essentially reconstituted and any structure that may exist in intact samples of natural soil will have been removed. The critical state strength parameters are therefore intrinsic to the particular soil and must depend only on the nature of the soil (i.e. on the grading and on the mineralogy, shape and texture of the grains). The tests used to describe the nature of soil are the grading tests and the Atterberg limits tests described in Chapters 5 and 7, and it is reasonable to suppose that the intrinsic properties will be related to these classification parameters.

Figure 9.16 Relationships between undrained strength, classification and water content in clays.

(a) Atterberg Limits, Undrained Strength and Water Content

The Atterberg limits described in Sec. 7.3 are for plastic clays. They measure the liquid limit (LL) and the plastic limit (PL), which are the water contents at which the undrained strength s_u is about 1.5 and 150 kPa respectively. As shown in Sec. 9.4, the voids ratio or the water content is linearly related to log s_u (see Fig. 9.5d) and the critical state line for a particular soil is shown in Fig. 9.16(a). The water content of soil can be expressed as the liquidity index (LI) given by

$$LI = \frac{w - PL}{PI} \tag{9.18}$$

where PI ($= LL - PL$) is the plasticity index. (At the liquid limit $LI = 1$ and at the plastic limit $LI = 0$.) Figure 9.16(b) shows the relationship between the liquidity index and undrained strength, and this holds for all plastic soils for which the Atterberg limits and undrained strengths can be measured. Thus, approximate estimates for the critical state undrained strength of soil can be obtained from measurements of the water content and Atterberg limits.

(b) Compressibility (C_c or λ)

From Fig. 9.4(c) the gradient of the critical state line is C_c, and this is also the gradient of the normal compression line. Total and effective shear stresses are the same and so the undrained strength s_u is given by

$$s_u = \tau_f' = \sigma_f' \tan \phi_c' \tag{9.19}$$

From Eq. (9.3),

$$e_{LL} = e_\Gamma - C_c \log\left(\frac{1.5}{\tan \phi_c'}\right) \tag{9.20}$$

$$e_{PL} = e_\Gamma - C_c \log\left(\frac{150}{\tan \phi_c'}\right) \tag{9.21}$$

and

$$e_{LL} - e_{PL} = C_c \log 100 = 2C_c \tag{9.22}$$

where e_{LL} and e_{PL} are the voids ratios at the liquid and plastic limits respectively. Since $e = wG_s$ we have

$$e_{LL} - e_{PL} = \frac{PI \times G_s}{100} \tag{9.23}$$

and

$$C_c = \frac{PI \times G_s}{200} \tag{9.24}$$

(Note that w is given as a decimal while liquid and plastic limits are given as percentages.)

Figure 9.17 Location of the critical state line through the Ω point.

Alternatively, from Eq. (9.10), and proceeding as above, we have

$$\lambda = \frac{\text{PI} \times G_s}{100 \ln 100} = \frac{\text{PI} \times G_s}{460} \tag{9.25}$$

Thus, estimates for the intrinsic compressibility of soils can be obtained from measurements of the Atterberg limits.

(c) Position of the Critical State Line (Γ or e_Γ)

It turns out that the critical state lines for many different soils pass through the same point, called the Ω (omega) point, shown in Fig. 9.17. The approximate coordinates of the Ω point, given by Schofield and Wroth (1968), are $v_\Omega = 1.25$ and $p'_\Omega = 1500\,\text{lb/in}^2 = 10\,\text{MPa}$; taking $K_0 = 0.5$, these correspond to $e_\Omega = 0.25$ and $\sigma'_\Omega = 15\,\text{MPa}$. From Eq. (9.3),

$$e_\Gamma = 0.25 + C_c \log 15\,000 \tag{9.26}$$

and, from Eq. (9.10),

$$\Gamma = 1.25 + \lambda \ln 10\,000 \tag{9.27}$$

Notice that C_c (or λ) describes the compressibility of soil while PI is the range of water content over which a plastic clay behaves as a soil (i.e. between the liquid and brittle states). It is not surprising that compressibility and plasticity index should be linearly related as in Eqs (9.24) and (9.25). Values for the parameters λ and Γ measured in isotropic compression tests on reconstituted samples of some typical plastic clays are given in Table 9.1.

(d) Critical Friction Angle

The critical friction angle for soil depends on the nature of the soil; some typical values are given in Table 9.1. (These relate to triaxial compression tests; values for triaxial extension and plane strain are usually only slightly larger.) For fine-grained soils ϕ'_c increases with decreasing plasticity. For coarse-grained soils ϕ'_c seems to depend mostly on the shape and roughness of the grains and on whether the soil is poorly graded or well graded. Be careful with these correlations as there are a number of cases where they do not apply. Notice that the Atterberg limits are measured on only the fine fraction of a soil so if the whole soil contains only a small percentage of fines the friction angle is likely to be higher than that given by the PI. A soil that has a high organic content may have a high PI but a relatively large friction angle (e.g. Bothkennar soil from the Firth of Forth in Scotland has PI ≈ 40 and $\phi'_c = 34°$).

(e) Swelling and Recompression

In the simple theories for soils described in Chapter 8 elastic compression and swelling for states inside the state boundary surface are given by Eqs (8.3) or (8.11) and the gradients of the swelling

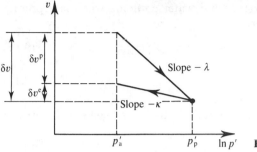

Figure 9.18 Elastic and plastic volume changes.

and recompression lines are κ or C_s for isotropic or one-dimensional loading respectively. (In Chapter 13 we will see, however, that soil is highly non-linear over a very wide range of loading so these parameters may not be so important.) Values of κ (or C_s) for soils turn out to be highly varied and seem to depend on what is happening to soil grains as they are loaded and unloaded.

An important parameter describing soil behaviour is the elastic volumetric strain ratio $\delta v^e/\delta v$. From Fig. 9.18, this is given by

$$\frac{\delta v^e}{\delta v} = \frac{\kappa}{\lambda} \tag{9.28}$$

For plastic clays values of κ/λ are generally in the range 0.2 to 0.5 depending on the PI; the larger values of κ/λ are associated with clays with a higher PI. This suggests that in clays there is some elastic distortion of the clay grains which is recovered on unloading. For coarse-grained soils the values of κ/λ are smaller, principally because the mechanism of compression includes fracturing of the grains which is not recovered on unloading.

Values for intrinsic parameters M, λ and Γ (or ϕ'_c, C_c and e_Γ) describing the critical state and normal compression lines for plastic clays are given in Table 9.1. These have been assembled from tests carried out at City University. They represent typical values, but these may vary due to differences in grading and mineralogy from sample to sample. The river sand, the carbonate sand and the decomposed granite have relatively large values of λ (i.e. they are relatively compressible), but it may be necessary to apply very large stresses to investigate the complete range of behaviour. Compressibility in coarse-grained soils is associated with particle fracturing and changes of grading, which has been demonstrated by observing the changes of grading after compression and shearing of carbonate sand and decomposed granite, both of which consist of weak and friable grains.

9.11 SUMMARY

1. During shearing soils ultimately reach a critical state where they continue to distort with no change of state (i.e. at constant shear stress, constant effective normal stress and constant water content).
2. Before the critical state there may be a peak state and after large strains clay soils reach a residual state. The peak state is associated with dilation and the residual state is associated with laminar flow.
3. The critical states of soils measured in triaxial tests are given by

$$q'_f = Mp'_f \tag{9.9}$$

$$v_f = \Gamma - \lambda \ln p'_f \tag{9.10}$$

where M, Γ and λ are constants for a particular soil. The undrained strength s_u is uniquely related to the water content, so for undrained loading (i.e. at constant water content) the undrained strength remains constant.

4. To take account of different effective normal stresses and water contents when interpreting soil test data the test results should be normalized with respect to p'_c or v_λ, given by

$$v_\lambda = v + \lambda \ln p' \qquad (9.11)$$

$$\ln p'_c = \frac{\Gamma - v}{\lambda} \qquad (9.12)$$

5. The critical states observed in triaxial tests are also found in shear tests, where the critical state lines are given by

$$\tau'_f = \sigma'_f \tan \phi'_c \qquad (9.2)$$

$$e_f = e_\Gamma - C_c \log \sigma'_f \qquad (9.3)$$

6. At the critical state soils are essentially perfectly frictional and the cohesion c' can be neglected.
7. The intrinsic critical state parameters M, λ and Γ (or ϕ'_c, C_c and e_Γ) depend principally on the nature of the soil and can often be estimated from the classification test parameters, particularly the Atterberg limits.

WORKED EXAMPLES

Example 9.1: Determination of critical state soil parameters A number of drained and undrained triaxial tests were carried out on normally consolidated and overconsolidated samples of the same soil. Table 9.2 gives values for the stress parameters q'_f and p'_f and the specific volume v_f when the samples had reached failure at their critical states.

Table 9.2

Test	p'_f (kPa)	q'_f (kPa)	v_f
A	600	588	1.82
B	285	280	1.97
C	400	390	1.90
D	256	250	1.99
E	150	146	2.10
F	200	195	2.04

The data are shown plotted in Fig. 9.19. Scaling from the diagram, $M = 0.98$, $\lambda = 0.20$. Substituting (say) $v = 1.82$ and $p' = 600\,\text{kPa}$ with $\lambda = 0.20$ into Eq. (9.10) we have $\Gamma = 3.10$.

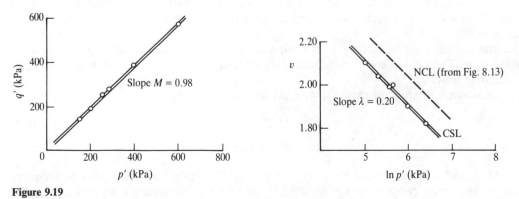

Figure 9.19

Example 9.2: Determination of critical states of soils A soil has the parameters $M = 0.98$, $\lambda = 0.20$ and $\Gamma = 3.10$. Four samples were isotropically compressed and swelled to the initial states shown in the first four columns of Table 9.3. In each case the pore pressure was $u_0 = 100\,\text{kPa}$. Each sample was tested by increasing q with the total mean stress p held constant: samples A and C were tested drained and samples B and D were tested undrained.

Table 9.3

Sample	p'_0 (kPa)	R_0	v_0	p'_f (kPa)	v_f	q'_f (kPa)	u_f (kPa)
A	600	1	1.97	600	1.82	588	100
B	600	1	1.97	284	1.97	278	416
C	150	4	2.04	150	2.09	147	100
D	150	4	2.04	200	2.04	196	50

For the drained tests $p'_f = p'_0$ and for the undrained tests $v_f = v_0$. From Eq. (9.10) the values of v_f in drained tests and p'_f in undrained tests are given by

$$v_f = \Gamma - \lambda \ln p'_f \qquad \text{or} \qquad p'_f = \exp\left(\frac{\Gamma - v_f}{\lambda}\right)$$

Notice that the tests were carried out with p constant so that $p_f = p_0$, the pore pressures at failure u_f in the undrained tests are given by

$$u_f = p_f - p'_f = p_0 - p'_f = p'_0 + u_0 - p'_f$$

From Eq. (9.9),

$$q_f = M p'_f$$

The points corresponding to isotropic compression and to failure at the critical state are shown in Fig. 9.20; these are linked by lines that represent approximately the state paths for the tests.

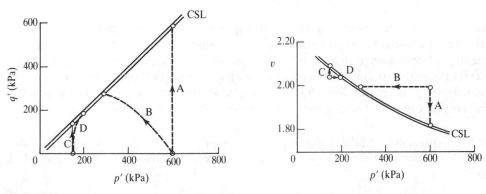

Figure 9.20

Example 9.3: Normalized critical states The initial states and the critical states given in Table 9.2 can be normalized with respect to the critical pressure p'_c given by Eq. (9.12) or with respect to the equivalent specific volume v_λ given by Eq. (9.11). The values for the normally consolidated samples A and C are given in Table 9.4 and the points representing the critical state and normal compression lines are given in Fig. 9.21.

Table 9.4

Sample	Initial state				Critical state						
	p'_0 (kPa)	p'_c (kPa)	p'_0/p'_c	v_λ	q'_f	p'_f (kPa)	p'_c (kPa)	q'_f/p'_c (kPa)	p'_f/p'_c	q'_f/p_f	v_λ
A	600	284	2.11	3.25	588	600	600	0.98	1.00	0.98	3.10
C	600	284	2.11	3.25	278	284	284	0.98	1.00	0.98	3.10

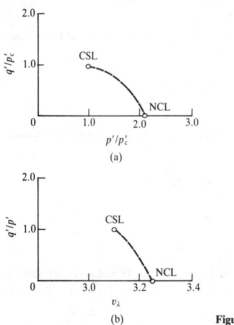

Figure 9.21

The initial and final state points may be joined together as shown by a line that represents approximately the state paths followed by the drained and undrained tests. Notice that in both tests the value of p'/p'_c decreases from 2.11 to 1.00. In the drained test this is because p'_c increases from 284 to 600 kPa as the specific volume decreases while $p' = 600$ kPa remains constant, but in the undrained test p' decreases from 600 to 284 kPa as the pore pressure increases while $p'_c = 284$ kPa remains constant because the specific volume does not change.

Example 9.4: Critical state Mohr circles and friction angle For the four tests given in Table 9.3 the principal stresses at the critical state can be calculated from Eqs (3.5) and (3.6). Rearranging:

$$\sigma'_a = p' + \tfrac{2}{3}q' \qquad \sigma'_r = p' - \tfrac{1}{3}q'$$

Table 9.5

Sample	q'_f (kPa)	p'_f (kPa)	σ'_{af} (kPa)	σ'_{rf} (kPa)
A	588	600	992	404
B	278	284	469	191
C	147	150	248	101
D	196	200	330	134

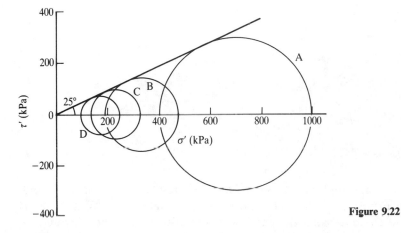

Figure 9.22

Values for σ'_a and σ'_r are given in Table 9.5 and the Mohr circles are given in Fig. 9.22. Scaling from the diagram, the critical friction angle is $\phi'_c = 25°$.

REFERENCES

Atkinson, J. H., J. A. Charles and H. K. Mhach (1990) 'Examination of erosion resistance of clays in embankment dams', *Quarterly Journal of Engineering Geology*, **23**, 103–108.

Skempton, A. W. (1964) 'Long term stability of clay slopes', *Geotechnique*, **14**, 77–101.

FURTHER READING

Atkinson, J. H. and P. L. Bransby (1978) *The Mechanics of Soils*, McGraw-Hill, London.

Muir Wood, D. M. (1991) *Soil Behaviour and Critical State Soil Mechanics*, Cambridge University Press, Cambridge.

Schofield, A. N. and C. P. Wroth (1968) *Critical State Soil Mechanics*, McGraw-Hill, London.

CHAPTER TEN

PEAK STATES

10.1 INTRODUCTION

Figure 10.1 shows the states of soil samples at the same effective stress σ' but at different voids ratios and overconsolidation ratios: at N the soil is normally consolidated, at W it is lightly overconsolidated or loose and the state is on the wet side of the critical state, and D_1 and D_2 are two states on the dry side where the soil is heavily overconsolidated or dense. For samples W and N on the wet side of critical the state parameters $\ln S_s$ and S_v (see Sec. 8.4) are positive and for samples D_1 and D_2 the state parameters are negative. Figure 10.2 shows the behaviour of these samples during drained shear tests and is similar to Fig. 9.1. At the critical states at C the samples have the same shear stress τ'_f, the same normal stress σ'_f and the same voids ratio e_f, but at the peak states the shear stresses and voids ratios are different. The idealized behaviour described in this chapter is based on experimental data given by Atkinson and Bransby (1978) and by Muir Wood (1991).

Peak states from shear tests on samples with different values of normal effective stress, overconsolidation ratio and voids ratio generally fall within the region OAB in Fig. 10.3 which is above the critical state line, and at first sight there is no clear relationship for the peak states as there was for the critical states. There are three ways of examining the peak states: the first is to make use of the Mohr–Coulomb equation with an apparent cohesion intercept, the second is to fit a curved line to the peak state points and the third is to include a contribution to strength from dilation.

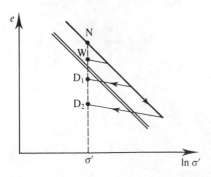

Figure 10.1 Initial states of samples at the same stress but different voids ratios.

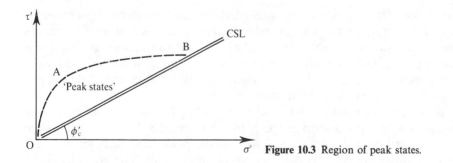

(a)

(b)

(c)   **Figure 10.2** Behaviour of samples in drained shear tests.

Figure 10.3 Region of peak states.

10.2 MOHR–COULOMB LINE IN SHEAR TESTS

Figure 10.4 shows peak states of two sets of samples which reached their peak states at voids ratios e_1 and e_2. These can be represented by the Mohr–Coulomb equation

$$\tau'_p = c'_{pe} + \sigma'_p \tan \phi'_p \qquad (10.1)$$

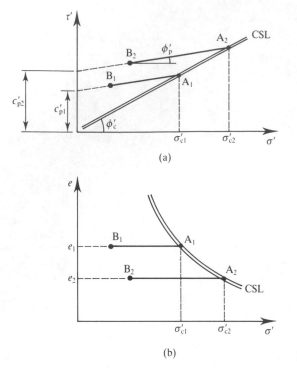

(a)

(b)

Figure 10.4 Peak states—cohesion.

where the subscripts p are there to make clear that Eq. (10.1) relates to the peak state and the subscript e in c'_{pe} is there because the cohesion intercept depends on the voids ratio.

There are a number of important things to notice about the peak states shown in Fig. 10.4. The peak friction angle ϕ'_p is less than the critical friction angle ϕ'_c and the peak state lines meet the critical state line at points such as A_1 and A_2. For any states to the right of the critical stateline in Fig. 10.4(b) the soil is on the wet side of critical and, on shearing, it compresses and reaches its critical state without a peak, as in Fig. 10.2. As a consequence peak states are associated with dense or overconsolidated soils on the dry side which dilate on shearing.

In Fig. 10.4 the peak state lines have been terminated at low stresses at points such as B_1 and B_2 and peak states at low stresses are not given by Eq. (10.1) (see Sec. 10.4). This means that the cohesion intercept c'_{pe} is not the shear stress which the soil can sustain at zero stress and it is merely a parameter required to define the Mohr–Coulomb equation. Since, in this case, these peak states apply equally for clean sand and reconstituted clays this cohesion should not be associated with cementing or interparticle attraction in clays.

In order to take account of the different voids ratios e_1 and e_2 in Fig. 10.4, we can make use of the normalizing parameter σ'_c described in Chapter 9. Figure 10.5 shows the peak state lines from Fig. 10.4 normalized with respect to σ'_c. Now all the peak state lines for different voids ratios reduce to the single line BA and A is the critical state point. The gradient of the peak state line is ϕ'_p and the dimensionless cohesion intercept is c'_p, which is given by

$$c'_p = \frac{c'_{pe}}{\sigma'_c} \tag{10.2}$$

From the geometry of Fig. 10.5, $c'_p = \tan \phi'_c - \tan \phi'_p$ so that the Mohr–Coulomb peak state parameters are not independent. From Eqs (9.7) and (10.2) the peak cohesion intercept c'_{pe} is given by

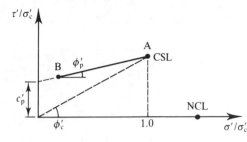

Figure 10.5 Normalized peak and critical states for shear tests.

$$\log\left(\frac{c'_{ce}}{c'_p}\right) = \frac{e_\Gamma - e}{C_c} \tag{10.3}$$

and so c'_{pe} increases with decreasing voids ratio.

10.3 MOHR–COULOMB LINE IN TRIAXIAL TESTS

In triaxial tests the peak states depend on the specific volume in the same way as for shear tests. Figure 10.6(a) shows the peak state line for the particular specific volume v in Fig. 10.6(b). In the region AB this is given by

$$q'_p = G_{pv} + H_p p'_p \tag{10.4}$$

where H_p is the gradient and G_{pv} is the intercept on the q' axis. The broken line OT at a gradient $dq'/dp' = 3$ represents the condition $\sigma'_r = 0$. Since uncemented soils cannot sustain tensile (negative) effective stresses, this represents a limit to possible states; the line OT is known as the tension cut-off and it is equivalent to the τ' axis for shear tests in Fig. 10.4. The parameter G_{pv} is simply a parameter that defines the position of the peak state line and is not necessarily the peak state at low effective stress.

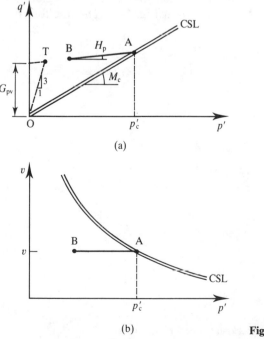

(a)

(b)

Figure 10.6 Peak states in triaxial tests.

Figure 10.7 Normalized peak and critical states for triaxial tests.

After normalization with respect to p'_c the results of triaxial tests appear as shown in Fig. 10.7, which is similar to Fig. 10.5 for shear tests. The critical state and normal compression lines reduce to single points and the peak states fall on a single line given by

$$\frac{q'_p}{p'_c} = G_p + H_p\left(\frac{p'_p}{p'_c}\right) \tag{10.5}$$

where the gradient is H_p and the cohesion intercept is G_p. The relationships between H_p and ϕ'_p are similar to those between M and ϕ'_c given by Eqs (9.16) and (9.17), and it is necessary to distinguish between values of H_p for compression and extension. From Figs 10.5 and 10.7, at very small effective stresses as σ'/σ'_c and p'/p'_c approach zero, and noting that $q'/p'_c \approx 2\tau'/\sigma'_c$, we have $G_p \approx 2c'_p$.

10.4 CURVED PEAK STATE LINES

At effective stresses lower than those normally applied in routine soil tests (i.e. at high overconsolidation ratios) the peak state line is markedly curved towards the origin, as illustrated in Fig. 10.8(a). For many soils it is only slightly curved at higher stresses in the region BA, where the line is close to the linear Mohr–Coulomb line shown in Fig. 10.5. The line for peak

(a)

(b)

Figure 10.8 Curved peak state line for shear tests.

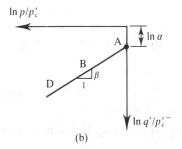

(a)

(b)

Figure 10.9 Curved peak state line for triaxial tests.

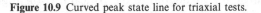

states normalized with respect to the critical stress σ'_c in Fig. 10.8(a) is similar to the broken line sketched in Fig. 9.7(a). The curved line in Fig. 10.8(a) is given by

$$\frac{\tau'_p}{\sigma'_c} = A\left(\frac{\sigma'_p}{\sigma'_c}\right)^b \tag{10.6}$$

where A and b are material properties (like c'_p and ϕ'_p). Equation (10.6) may be written as

$$\ln\left(\frac{\tau'_p}{\sigma'_c}\right) = \ln A + b \ln\left(\frac{\sigma'_p}{\sigma'_c}\right) \tag{10.7}$$

This is shown in Fig. 10.8(b) and Eq. (10.7) provides a convenient method for determining values for the parameters A and b. Figure 10.9(a) shows peak states from triaxial tests normalized with respect to p'_c and represented as a curved line and Fig. 10.9(b) is the same behaviour plotted with logarithmic scales. The peak states in Fig. 10.9 are given by

$$\ln\left(\frac{q'_p}{p'_c}\right) = \ln \alpha + \beta \ln\left(\frac{p'_p}{p'_c}\right) \tag{10.8}$$

where α and β are soil parameters. The curved peak state lines in Figs 10.8 and 10.9 both end at the critical state point where the peak and critical states are the same. Hence $A = \tan \phi'_c$ and $\alpha = M$, so that only the parameters b or β are required to define the curved peak state envelope.

10.5 PEAK STATES AND DILATION

An alternative approach to understanding the peak states of soils is to recognize that during shearing of a dilating soil the shear stresses must both overcome friction between the grains and lift the normal loads. For a given normal stress the peak shear stresses will increase with increasing rate of dilation.

Figure 10.10 A model for shearing and dilation.

Figure 10.10(a) shows the stresses and displacements for an increment of displacement of a shear test and Fig. 10.10(b) shows horizontal and vertical forces on a frictional block on an inclined plane. The mechanics of both are similar and so the relationships between τ' and σ' in the shear test will be analogous to the relationships between T and N and the slope angle i is analogous to ψ. If $i = 0$ the block slides when

$$\frac{T}{N} = \tan \mu \tag{10.9}$$

so the friction angle μ is analogous to the critical friction angle ϕ'_c. From Fig. 10.10(b), resolving horizontally and vertically and after some algebra, we have

$$\frac{T}{N} = \tan(\mu + i) \tag{10.10}$$

(Readers should do the simple mechanics and algebra themselves.) Following the analogy between the shear test and the sliding block, the behaviour of soil can be represented by

$$\frac{\tau'}{\sigma'} = \tan(\phi'_c + \psi) \tag{10.11}$$

Figure 10.11 shows the behaviour of overconsolidated or dense soil on the dry side of critical

Figure 10.11 Shearing and dilation in shear tests.

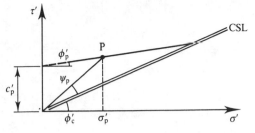

Figure 10.12 Peak state of a dilating soil.

in a drained shear test and it is essentially the same as Fig. 9.1. There are two points, A and C, where $\psi = 0$ and so, from Eq. (10.11), $\tau'/\sigma' = \tan \phi'_c$ and at the peak state at P the value of ψ_p is the maximum.

Figure 10.12 shows a point P representing the peak state of a soil at a particular stress σ'_p and voids ratio e_p, and it lies on the appropriate Mohr–Coulomb line given by c'_p and ϕ'_p. It also lies on the line given by Eq. (10.11) with a critical friction angle ϕ'_c and a peak dilation angle ψ_p. Figure 10.13 shows four peak state points: B_1 and C_1 have the same voids ratio e_1 and they lie on the same Mohr–Coulomb line given by c'_{p1} and ϕ'_p. (B_2 and C_2 are similar points at the same voids ratio e_2 and they lie on the Mohr–Coulomb line given by c'_{p2} and ϕ'_p.) Points B_1 and B_2 have the same normal stress σ'_b, but B_2 is more heavily overconsolidated than B_1 and has a lower voids ratio. Since B_2 and B_1 will reach the same critical states at B_c, sample B_2 must dilate more (i.e. have a larger value of ψ_p) than sample B_1. (C_1 and C_2 are similar points.) Points B_1 and C_2 have the same overconsolidation ratio R_0 but different voids ratios and normal stresses; their peak states are given by Eq. (10.11) with the same value of ψ_p.

For triaxial tests the equation that is analogous to Eq. (10.11) is

$$\frac{q'}{p'} = M - \frac{d\varepsilon_v}{d\varepsilon_s} \tag{10.12}$$

Figure 10.13 Peak states of soils with different states.

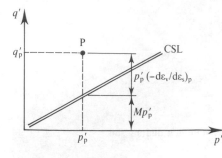

Figure 10.14 Peak state of dilating soil.

(The negative sign is required because $d\varepsilon_v$ is negative for dilation.) Equation (10.12) shows that the stress ratio q'/p' is the sum of the critical stress ratio M and the rate of dilation $-d\varepsilon_v/d\varepsilon_s$. Figure 10.14 shows a peak point P at a particular stress. The peak deviator stress given by

$$q'_p = Mp'_p - p'_p\left(\frac{d\varepsilon_v}{d\varepsilon_s}\right)_p \tag{10.13}$$

is made up of contributions from the friction parameter M and the rate of dilation.

10.6 VARIATION OF PEAK STATE WITH INITIAL STATE

Equations (10.11) and (10.12) demonstrate that the peak stress ratio increases with the peak rate of dilation ψ_p or $(d\varepsilon_v/d\varepsilon_s)_p$, and we would expect to find that the peak rate of dilation depends on the initial state or overconsolidation ratio. (In Chapter 11 we will see that these equations apply for states other than the peak but, for the present, I will restrict the discussion to the peak states.) Figure 10.15 shows two samples of the same soil, 1 and 2, at the same stress p'_0 but with different specific volumes, v_1 and v_2. The two soils have different values of v_λ and different overconsolidation ratios; soil 2 is the more heavily overconsolidated. During drained shearing at constant p' the soils dilate and both ultimately reach the same critical state at C. Figure 10.15(b) and (d) shows the corresponding changes of specific volume and the volumetric strains. The points P_1 and P_2 are the maximum gradients and ψ_1 and ψ_2 are the maximum rates of dilation; it is obvious that ψ_2 is greater than ψ_1 and so the peak angle of dilation increases with increasing overconsolidation ratio (i.e. with decreasing v_λ). Figure 10.16(a) shows how the peak state $(q'/p')_p$ varies with v_λ (and hence with overconsolidation); at the critical state line $v_\lambda = \Gamma$ and the rate of dilation is zero. Figure 10.16(b) illustrates the corresponding variation of the peak stress ratio $(\tau'/\sigma')_p$ with e_λ. The relationships in Fig. 10.16 correspond to the broken lines sketched in Figs 9.7(b) and 9.10(b).

Peak states may also be interpreted in terms of the state parameters S_v or S_s. (Note that these are related by $S_v = \lambda \ln S_s$.) Figure 10.17(a) shows two states 1 and 2 similar to those in Fig. 10.16(a) and the state parameters are S_{v1} and S_{v2}. Figure 10.17(b) illustrates the variation of peak stress ratio with state parameter; notice that S_v is negative for states on the dry side of critical and at the critical state $S_v = 0$ and $q'/p' = M$.

In Fig. 10.16 the values of v_λ and e_λ depend on both the nature and the state of the soil. For fine-grained soils it is possible to normalize these further by relating them to the Atterberg limits. Figure 9.16 showed the relationships between undrained strength and liquidity index LI given by Eq. (5.10) and, noting that s_u/p'_f is constant for a given soil, this can be redrawn as Fig. 10.18(a). (For a typical clay soil $s_u/p'_f \approx 0.5$ and so the reference strength $s_u = 1.5$ kPa in Fig. 9.16 corresponding to the liquid limit should be replaced by $p' = 3$ kPa in Fig. 10.18.) The current state at A can be represented by an equivalent liquidity index, LI_λ, as shown (Schofield, 1980).

Figure 10.15 Dilation and peak states related to initial states.

Figure 10.16 Variation of peak stress ratio with peak volume state.

Figure 10.17 Relationship between peak stress ratio and state parameter.

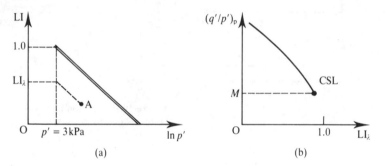

Figure 10.18 Stress ratio and dilatancy related to liquidity index.

The relationship between the peak stress ratio and LI_λ in Fig. 10.18(b) would then apply for all plastic soils for which it was possible to determine the Atterberg limits. (Notice that for a state on the wet side of critical, $LI_\lambda > 1$.)

10.7 SUMMARY

1. Overconsolidated soils on the dry side of critical generally reach peak states where the shear stress and the stress ratio are larger than those at the critical state.
2. At a particular normal stress the peak state depends on the voids ratio but, after normalization with respect to σ'_c or p'_c, the peak states are the same.
3. The peak states can be represented by either of the following:

   (a) a linear Mohr–Coulomb line,
   (b) a curved envelope, or
   (c) a peak stress ratio and angle of dilation.

4. The peak state is governed particularly by the initial overconsolidation ratio or by the state parameter.

WORKED EXAMPLES

Example 10.1: Determination of peak state parameters Table 10.1 shows data obtained at the peak state from a series of shear tests on the same soil as that in the examples in Chapter 9. (Note that the samples reached their peak state at one of only two different voids ratios.) The peak states are plotted in Fig. 10.19. The points fall close to two straight lines given by $c'_{pe} = 60$ kPa and $c'_{pe} = 130$ kPa with $\phi'_p = 15°$ in both cases.

Table 10.1

Sample	τ'_p (kPa)	σ'_p (kPa)	e_p	σ'_c (kPa)	τ'_p/σ'_c	σ'_p/σ'_c
A	138	300	1.03	300	0.46	1.00
C	123	240	1.03	300	0.41	0.80
E	108	180	1.03	300	0.36	0.60
G	93	120	1.03	300	0.31	0.40
B	264	540	0.89	606	0.44	0.90
D	228	420	0.89	606	0.38	0.70
F	198	300	0.89	606	0.33	0.50

Figure 10.19

The test results can be normalized with respect to the equivalent stress σ'_c given by Eq. (9.7). The soil parameters are $C_c = 0.46$ and $e_\Gamma = 2.17$. The normalized stresses are given in Table 10.1 and these are plotted in Fig. 10.20. The data now all fall close to a single straight line given by $c'_p = 0.2$ and $\phi'_p = 15°$.

Figure 10.20

Example 10.2: Curved peak state envelope   A further set of four shear tests was carried out in addition to those described in Example 10.1 and the results are given in Table 10.2. The peak states are plotted in Fig. 10.21 together with the data from Table 10.1. The points fall close to

Table 10.2

Sample	τ'_p (kPa)	σ'_p (kPa)	e_p	σ'_c (kPa)	τ'_p/σ'_c	σ'_p/σ'_c
J	63	60	1.03	300	0.21	0.20
L	30	15	1.03	300	0.10	0.05
H	156	180	0.89	606	0.26	0.30
K	84	60	0.89	606	0.14	0.10

Figure 10.21

two curved lines, although at large stresses, for which the data are those given in Table 10.1, the lines are very nearly straight. As before, the data can be normalized with respect to the equivalent stress σ'_c. The normalized stresses are given in Table 10.2 and plotted in Fig. 10.22(a). The data now all fall close to a single curved line. The data are plotted to logarithmic scales in Fig. 10.22(b). The gradient of the line is b in Eq. (10.7) and, scaling from the diagram, $b = 0.5$.

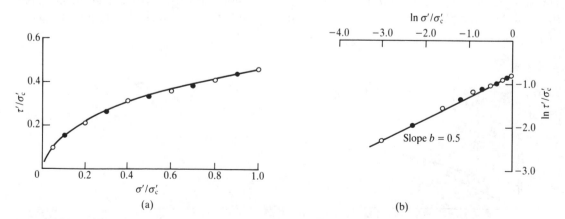

Figure 10.22

Example 10.3: Dilation and equivalent state   Table 10.3 shows the data for the peak states for the set of shear tests given in Tables 10.1 and 10.2. The critical friction angle is ϕ'_c and hence the angle of dilation ψ is given by Eq. (10.11). Values for e_λ are calculated from Eq. (9.6) and Fig. 10.23 shows the variation of $\phi' + \psi$ with e_λ (see Fig. 10.16).

Table 10.3

Sample	τ'_p (kPa)	σ'_p (kPa)	e_p	$\phi'_c + \psi$	ψ	e_λ
A	138	300	1.03	24.7	−0.3	2.17
B	264	540	0.89	26.1	1.1	2.15
C	123	240	1.03	27.1	2.1	2.12
D	228	420	0.89	28.5	3.5	2.10
E	108	180	1.03	31.0	6.0	2.07
F	198	300	0.89	33.4	8.4	2.03
G	93	120	1.03	37.8	12.8	1.99
H	156	180	0.89	40.9	15.9	1.93
J	63	60	1.03	46.4	21.4	1.85
K	84	60	0.89	54.5	29.5	1.71
L	30	15	1.03	63.4	38.4	1.57

Figure 10.23

REFERENCE

Schofield, A. N. (1980) 'Cambridge geotechnical centrifuge operations', *Geotechnique*, **30**, 3, 227–268.

FURTHER READING

Atkinson, J. H. and P. L. Bransby (1978) *The Mechanics of Soils*, McGraw-Hill, London.
Muir Wood, D. M. (1991) *Soil Behaviour and Critical State Soil Mechanics*, Cambridge University Press, Cambridge.
Schofield, A. N. and C. P. Wroth (1968) *Critical State Soil Mechanics*, McGraw-Hill, London.
Taylor, D. W. (1948) *Fundamentals of Soil Mechanics*, Wiley, New York.

CHAPTER ELEVEN

BEHAVIOUR OF SOIL BEFORE FAILURE

11.1 INTRODUCTION

In laboratory triaxial and shear tests, and in the ground, soil is loaded from some initial state and will ultimately reach some critical state. Initial states for isotropic and one-dimensional compression and swelling were discussed in Chapter 8; knowing the history of loading and unloading, the initial specific volume and overconsolidation ratio are fixed. Critical states were discussed in Chapter 9; knowing either the mean normal stress or the specific volume at the critical state, the critical state strength is fixed. This means that we can generally calculate initial and critical states for any drained or undrained loading and it is now necessary to consider how the states change from the initial to the critical.

We already have some information about these intermediate states. In Chapter 10 we considered states corresponding to the peak stress ratio of samples with specific volumes or voids ratios initially lower than the critical state and which dilated during drained shearing. Also, in Chapter 9, I suggested that there might be unique states between the normal compression and critical state lines as shown in Figs 9.7 and 9.10.

11.2 WET SIDE AND DRY SIDE OF CRITICAL

During drained shearing soil may either compress or dilate, as illustrated in Figs 9.1 and 10.2, and during undrained shearing pore pressures may either increase or decrease. What actually happens depends on the position of the initial state with respect to the critical state line. We can now see the significance of the distinction made in Sec. 8.4 between states on the wet side of the critical state (i.e. normally consolidated or lightly overconsolidated clays or loose sands) and states on the dry side (i.e. heavily overconsolidated clays or dense sands).

Figures 11.1 and 11.2 illustrate the idealized behaviour of soils initially on the wet side or on the dry side during undrained or drained triaxial tests. In Fig. 11.1 the state at W is normally consolidated (i.e. on the wet side) and the state at D is heavily overconsolidated (i.e. on the dry side), both having the same initial specific volume. During any shearing test the states must move towards, and ultimately reach, the critical state line. For undrained loading the states must

(a)

(b)

Figure 11.1 Behaviour of soil during undrained shearing.

(a)

(b)

Figure 11.2 Behaviour of soil during drained constant p' shearing.

remain at constant volume and both samples reach the critical state line at F_u, where they have the same undrained strength because they have the same specific volume. The total stress path $I \rightarrow F$, corresponds to a constant mean total stress p and the horizontal distances between the total and effective stress paths are equal to the pore pressure. (A test with constant p can be carried out in a hydraulic triaxial cell (see Chapter 7) by simultaneously reducing the cell pressure and increasing the axial stress in the ratio $\delta\sigma_a = -2\delta\sigma_r$.) You can see from Fig. 11.1

Figure 11.3 States on the wet side and the dry side of critical.

that for the soil initially on the wet side the pore pressure increases on shearing, while for the soil initially on the dry side the pore pressure reduces. Notice that the overconsolidated soil reached a peak stress ratio at P but the deviator stress at F_u is greater than that at P. Notice also that the loading path D → P for the overconsolidated soil before the peak state is linear and vertical (i.e. $\delta p' = 0$).

Figure 11.2 shows the same two initial states but with the paths for drained shearing with constant p'. Again both paths must move towards, and ultimately reach, the critical state line. The soil initially on the wet side compresses on shearing and ultimately fails at F_w. The soil initially on the dry side first shears at constant volume to the peak state at P but then it dilates and the shear stress reduces as the specific volume increases. The shear stresses at the failure points, F_w and F_d, are different because the effective stresses and specific volumes are different.

The principal distinction between soils that compress on drained shearing or where pore pressures increase on undrained shearing and soils that dilate or where pore pressures decrease is in whether the initial state lies to the right (i.e. on the wet side) or to the left (i.e. on the dry side) of the critical state line as illustrated in Fig. 11.3. Soils initially on the wet side compress during drained shearing or the pore pressures increase during undrained shearing, while soils initially on the dry side dilate or pore pressures decrease.

The distinction between the dry side and the wet side of critical is very important in determining the basic characteristics of soil behaviour. Soils must be heavily overconsolidated (R_p about 3) to be on the dry side while soils that are normally or lightly overconsolidated ($R_p <$ about 2) will be on the wet side. Remember that the initial state could also be described by the state parameters S_v or S_s: for states on the wet side S_v and $\ln S_s$ are positive and on the dry side they are negative.

11.3 STATE BOUNDARY SURFACE FOR SOIL

We have already found cases where the possible states of soils were limited; these are shown in Fig. 11.4. As discussed in Sec. 8.3 and illustrated in Fig. 8.4, the isotropic normal compression line represents a boundary to all possible states of isotropic compression; the state can move below (i.e. inside) the boundary by unloading, but it cannot move outside the normal compression line. Similarly, the peak envelope must represent a boundary to all possible states since, by definition, this represents the limiting or peak states. Remember that the peak state line in Fig. 11.4(a) corresponds to one specific volume. There will be other peak state lines corresponding to other specific volumes and together these will form a peak state surface. The surface can be reduced to a line by normalization as described in Sec. 9.5. The peak state boundary surface, normalized with respect to the critical pressure p'_c, is shown in Fig. 10.9.

The peak state surface is a boundary on the dry side of critical and it is now necessary to examine whether there is a well-defined state boundary on the wet side. If there is it will join

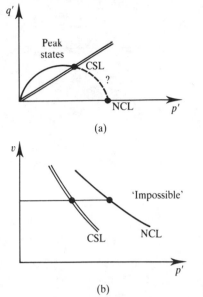

(a)

(b)

Figure 11.4 Part of a state boundary surface on the dry side.

the isotropic normal compression and critical state lines and it might look like the broken line in Fig. 11.4.

Figure 11.5 shows paths for three different initial states all on the wet side of critical. P and V are on the isotropic normal compression line; P is sheared drained with p' constant and V is sheared undrained and the paths cross at S. R is initially anisotropically compressed and it is compressed further at a constant stress ratio $q'/p' = \eta'$ so that the state passes through the point S. (Notice that the normal compression line is like this path but with $\eta' = 0$ and so is the critical state line but with $\eta' = M$.) We can easily arrange for all the stress paths in Fig. 11.5(a) to pass through the same point S, but the question is whether they all have the same specific volume at S in Fig. 11.5(b).

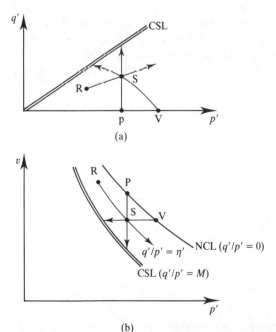

(a)

(b)

Figure 11.5 State paths for normally consoldated soil.

(a)

(b)

Figure 11.6 Part of a state boundary surface on the wet side.

The best way to examine this is to normalize the states with respect to the critical pressure p'_c or with respect to the equivalent volume v_λ. The resulting normalized state boundary surface is shown in Fig. 11.6. As before, the critical state and isotropic normal compression lines reduce to single points and the anisotropic compression line RS reduces to a single point S. Also shown in Fig. 11.6 are the parts of the state boundary surface on the dry side of critical, corresponding to the peak states, from Figs. 10.9 and 10.16.

The state boundary surface in Fig. 11.6 has been drawn as a smooth curve linking the wet side and the dry side. Later, in Chapter 12, this will be represented by a simple mathematical expression. Do not forget that the line shown in Fig. 11.6, which has normalized axes, is really a three-dimensional surface in the set of axes $q':p':v$. This surface is rather difficult to draw, which is why it is easier to normalize the results first. Figure 11.7 illustrates the three-dimensional

Figure 11.7 A state boundary surface for soil.

surface; this shows constant specific volume sections as full lines and constant stress ratio (i.e. constant η') sections as broken lines. Data from soil tests that demonstrate the existence of unique state boundary surfaces were given by Atkinson and Bransby (1978) and by Muir Wood (1991). The part of the state boundary surface on the wet side of critical (i.e. between the normal compression line and the critical state line) is sometimes known as the Roscoe surface and the part on the dry side corresponding to peak states is sometimes known as the Hvorslev surface (Atkinson and Bransby, 1978).

11.4 ELASTIC BEHAVIOUR AT STATES INSIDE THE STATE BOUNDARY SURFACE

The state boundary surface is a boundary to all possible states of a reconstituted soil. The state cannot exist outside the surface—by definition—although later we will find cases of cemented soils where unstable states outside the boundary surface for reconstituted soil can occur. If soil with a state on the surface is unloaded the state moves inside the surface and, on reloading the state, will move back to, but not outside, the surface. Thus, the state boundary surface can also be a yield surface like that shown in Fig. 3.16. If the boundary surface is a yield surface then while the state is on the surface there are simultaneous elastic and plastic strains, but if the state is brought inside the boundary surface, by unloading, the strains are assumed to be purely elastic. This is a highly idealized model for soil behaviour and we now know that there are inelastic strains when the state is inside the boundary surface. These aspects of soil stress–strain behaviour will be considered briefly in Chapter 13.

The idealized behaviour of soil during isotropic compression and swelling was considered in Secs 8.2 and 8.3 (see Figs 8.2 to 8.6) and is illustrated in Fig. 11.8. This shows a sequence of loading and unloading from A to D where the overconsolidation ratios are the same but the specific volumes are different. Between B and C the state was on the normal compression line (i.e. on the state boundary surface) and the soil yielded and hardened as the yield stress increased by $\delta p_y'$ with an irrecoverable plastic volume change δv^p. Along AB and CD the state was inside the boundary surfaces and the behaviour is taken to be elastic.

The stress–strain behaviour of an isotropic elastic material is decoupled (i.e. the shearing and volumetric effects are separated) and from Eq. (3.21),

$$\delta \varepsilon_s = \frac{1}{3G'} \delta q' \tag{11.1}$$

$$\delta \varepsilon_v = \frac{1}{K'} \delta p' \tag{11.2}$$

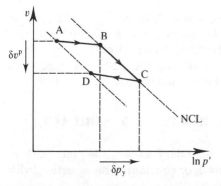

Figure 11.8 Elastic and plastic compression.

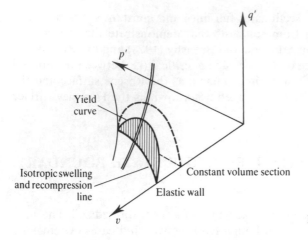

Figure 11.9 Elastic wall and yield curve.

Another expression for the elastic volumetric strains can be obtained from the equation for the swelling and recompression lines (see Sec. 8.2) as

$$\delta \varepsilon_v = \frac{\kappa}{v p'} \delta p' \tag{11.3}$$

where κ is the slope of the lines AB and CD in Fig. 11.8. A similar expression for shearing can be written as

$$\delta \varepsilon_s = \frac{g}{3 v p'} \delta q' \tag{11.4}$$

where g is a soil parameter which describes shear stiffness in the same way that κ describes volumetric stiffness. (The basic assumption made here is that $G'/K' = \kappa/g =$ constant, which implies that Poisson's ratio is a constant.)

With the simple idealization that soil is isotropic and elastic, shear and volumetric effects are decoupled and volume changes are related only to changes of p' and are independent of any change of q'. This means that, inside the state boundary surface, the state must remain on a vertical plane above a particular swelling and recompression line. This vertical plane is sometimes known as an elastic wall (see Fig. 11.9). Notice that an elastic wall is different from a constant volume section (except for the case of a soil with $\kappa = 0$). Since the soil yields when the state reaches the boundary surface a yield curve is the intersection of an elastic wall with the state boundary surface, as shown in Fig. 11.9. Remember that there will be an infinite number of elastic walls, each above a particular swelling and recompression line, and an infinite number of yield curves.

For any undrained loading path on an elastic wall $\delta \varepsilon_v = 0$ and, from Eq. (11.3), $\delta p' = 0$. Figure 11.10 shows the state paths for undrained shearing of lightly overconsolidated soil from W on the wet side of critical and of heavily overconsolidated soil from D on the dry side of critical, both with the same specific volume. For the initial loading the paths are linear and vertical ($\delta p' = 0$) and the soils yield at Y_W and Y_D when the states reach the boundary surface.

11.5 UNDRAINED LOADING ON THE STATE BOUNDARY SURFACE

Beyond the yield point where the state reaches the state boundary surface the state path in undrained loading must follow the intersection of an undrained, or constant volume section, with

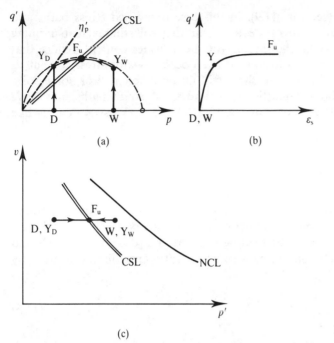

(a)

(b)

(c)

Figure 11.10 State paths for undrained loading.

the state boundary surface as shown in Fig. 11.9. Thus, in Fig. 11.10(a), the state paths are $Y_D \rightarrow F_u$ and $Y_W \rightarrow F_u$ and the stress–strain curves will be like those shown in Fig. 11.10(b). Notice that neither of the soils has any peak deviator stress q' but the soil on the dry side has a peak stress ratio η'_p at Y_D.

In Fig. 11.10 the undrained section of the state boundary surface has been drawn approximately symmetric about the critical state point F_u, but this is not usually the case and undrained sections will commonly be asymmetric with a peak on the dry side, as shown in Fig. 11.11(a). The undrained behaviour shown in Fig. 11.11(a) arises from the geometry of the state

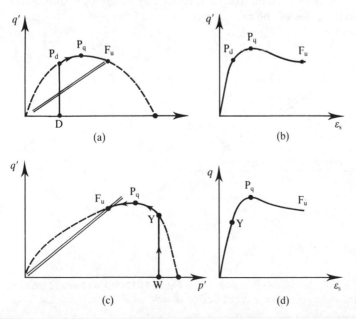

(a)

(b)

(c)

(d)

Figure 11.11 State paths for undrained loading.

boundary surface and the elastic wall (see Fig. 11.9). In this case undrained stress paths and stress–strain curves will have peak deviator stress states at P_q, but they will still reach an ultimate or critical state at F_u which depends only on the specific volume or water content. Notice that for soil on the dry side there is a peak stress ratio state at P_d, a peak deviator stress state at P_q and an ultimate or critical state at F_u; in general, these will all be different. For some soils, particularly natural lightly overconsolidated plastic clays, undrained stress paths may have peaks on the wet side of the critical state as shown in Fig. 11.11(c) and stress–strain curves like those shown in Fig. 11.11(d).

11.6 STRESS RATIO AND DILATION

Figure 11.12(a) and (b) shows the variations of stress ratio and volumetric strain with shear strain for an ideal soil sheared from states initially either on the wet side or on the dry side and are similar to Figs 9.1 and 10.2. In Fig. 11.12 the loading was terminated before the soils had reached their critical states at C. The relationship between stress ratio and dilation is given by Eq. (10.12) as

$$\frac{q'}{p'} = M - \frac{d\varepsilon_v}{d\varepsilon_s} \tag{11.5}$$

and in Sec. 10.5 and 10.6 I showed that this described the states of soils at their peak state.

Providing that elastic strains are relatively small compared to the plastic strains, Eq. (11.5) also applies to states before and after the peak and to soils on the wet side and on the dry side (except at states close to the start of the shearing where the behaviour is essentially elastic). Figure 11.12(c) shows Eq. (11.5) as q'/p' against $d\varepsilon_v/d\varepsilon_s$ for the normally consolidated soil and for the overconsolidated soil at states beyond A in Fig. 11.12(a). There are two points, A and C, where the rate of volume change is zero and $q'/p' = M$. Consequently, by plotting soil test data as q'/p' against $d\varepsilon_v/d\varepsilon_s$ the position of the critical state point C can be found even if the loading is terminated before the samples have reached their critical states. It is best to conduct tests on both normally consolidated and overconsolidated samples of clay or on loose and dense samples of sand to obtain data on both sides of the critical state.

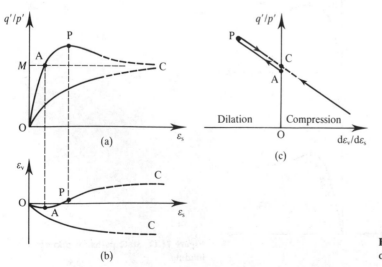

Figure 11.12 Stress ratio and dilation of soil.

11.7 SOFTENING OF SOIL BEYOND THE PEAK STATE AND DEVELOPMENT OF SLIP SURFACES

An important feature of straining of soils on the dry side of critical at relative large strains is the development of strong discontinuities, or slip surfaces. (Strains in slip surfaces were considered in Sec. 2.7.) You can see these if you squash cold butter, stiff clay or a sand-castle in unconfined compression, but you will not see slip surfaces if you squash soft butter, soft clay or loose dry sand.

What you see are not really slip surfaces but thin regions of intense shearing through material that is essentially rigid on either side. In soils these thin regions may be only a few grains thick, but as they have finite thickness the soil inside them can change in volume. This means that once slip surfaces have developed soil is no longer homogeneous and the shear and volumetric strains are highly non-uniform, so that measurements of strain made at the boundary of a test sample will not properly measure the strains and specific volume in the soil, which is straining in the slip surfaces. Consequently, once a slip surface has developed in soil samples you cannot rely on the conventional measurements of strain and specific volume.

Slip surfaces develop in materials that soften (i.e. the available shear stress reduces) on straining. This is because as soon as a slip surface starts to form the strains become non-uniform and there is additional straining in the thin region. The material then becomes weaker in the region of the larger strains so that more strain will accumulate and the slip surface will grow. Thus, in soils we would normally expect to find slip surfaces occurring mostly during shearing on the dry side of critical. Notice that the shear straining will be associated with dilation so soil within a slip surface will have a specific volume which is larger than that of the surrounding material.

On the other hand, a clay that is normally consolidated or lightly overconsolidated and on the wet side will compress on straining and so the water content will decrease as it strengthens and hardens. In this case, if there is any non-uniform straining the material in the region of the larger strains becomes stronger than the neighbouring material so a well-defined slip surface never develops. If you carefully examine samples of soft clay or loose sand after failure in triaxial tests you can often find many faint slip surfaces, but no strong discontinuities (it helps if you let clay samples dry). Similar faint lines can be found on the surface of metal specimens after straining; they are called Luders lines.

Slip surfaces, or strong strain discontinuities, do occur in soils as they suffer large strains and we will come across these later as we develop solutions for stability of foundations and slopes. For the present you must remember that if slip surfaces occur in laboratory specimens you can no longer trust the measurements of specific volume. Another reason why some people do not believe that soils have unique critical states is that they plot the average specific volume of the whole specimen without making any allowance for the local increase in specific volume in slip surfaces. This applies equally to drained and to undrained tests. In tests that are nominally undrained in the sense that the overall volume does not change, there can still be local drainage of water into slip surfaces from the surrounding soil.

11.8 SUMMARY

1. The initial state of soil, before shearing, is fixed by the appropriate normal compression and swelling lines and the final state is fixed by the critical state line. The path between the initial and final states is governed by the loading (i.e. drained or undrained) and by the state boundary surface.

2. There is an important distinction to be made between the behaviour of soils on the wet side of critical (which compress on drained loading, or where the pore pressures rise on undrained loading) and soils on the dry side of critical (which dilate on shearing, or where the pore pressures fall).

3. In the simple idealization the behaviour is taken to be elastic when the state is inside the state boundary surface. Yielding and plastic straining occur as the state moves on the state boundary surface.

4. There are relationships between stress ratio and dilation for states on the state boundary surface on the wet side and on the dry side of the critical state. These relationships provide a means of determining the critical state of soil from tests in which the sample did not reach the critical state.

5. Overconsolidated soils, on the dry side of critical, which soften on shearing beyond the peak often develop strong slip surfaces where intense shearing and volume changes are concentrated in a very thin region of material. In this case measurements made at the boundaries of a test sample become unreliable.

WORKED EXAMPLES

Example 11.1: Determination of state path and yielding A soil has the parameters $M = 0.98$, $\lambda = 0.20$, $\kappa = 0.05$ and $N = 3.25$. A constant volume section of the state boundary surface is a semi-circle passing through the origin. Samples were isotropically compressed and swelled in a stress path triaxial cell to different stresses but the same initial specific volume $v_0 = 1.97$; the initial stresses were: sample A, $p_0' = 600$ kPa, sample B, $p_0' = 400$ kPa, sample C, $p_0' = 150$ kPa (sample A was normally consolidated). The samples were tested undrained by increasing q with p held constant.

The state paths are shown in Fig. 11.13. When the state is inside the state boundary surface the behaviour is elastic and shearing and volumetric effects are decoupled; hence $\delta p' = 0$ for undrained loading. The states of the samples after compression and swelling, at their yield points and at failure at their critical states, shown in Table 11.1, were found by scaling from the diagram.

Figure 11.13

Table 11.1

Sample	Initial state p'_0 (kPa)	v_0	Yield point q'_y (kPa)	p'_y (kPa)	v_y	Critical state q'_f (kPa)	p'_f (kPa)	v_f
A	600	1.97	0	600	1.97	294	300	1.97
B	400	1.97	280	400	1.97	294	300	1.97
C	150	1.97	260	150	1.97	294	300	1.97

Example 11.2: Determination of state path and yielding Three further samples D, E and F of the soil described in Example 11.1 were prepared at the same initial state as samples A, B and C. Each sample was tested drained following a stress path with increasing q' with p' held constant.

The state paths are shown in Fig. 11.14. When the state is inside the state boundary surface the behaviour is elastic and shearing and volumetric effects are decoupled; hence $\delta v = 0$ for constant p' stress path tests. The states of the samples after compression and swelling, at their yield points and at failure at their critical states, shown in Table 11.2, were found by scaling from the diagram.

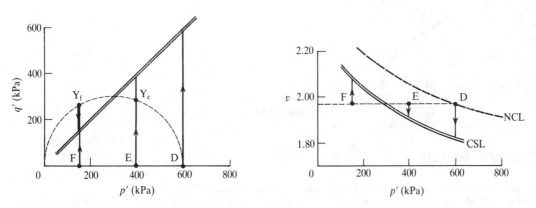

Figure 11.14

Table 11.2

Sample	Initial state p'_0 (kPa)	v_0	Yield point q'_y (kPa)	p'_y (kPa)	v_y	Critical state q'_f (kPa)	p'_f (kPa)	v_f
D	600	1.97	0	600	1.97	588	600	1.83
E	400	1.97	280	400	1.97	392	400	1.90
F	150	1.97	260	150	1.97	147	150	2.10

Example 11.3: Calculation of undrained stress path A soil has the parameters $M = 0.98$, $\lambda = 0.20$, $\kappa = 0.05$ and $N = 3.25$, but the shape of the state boundary surface is unknown. A sample is isotropically normally compressed in a triaxial apparatus to $p_0' = 600\,\text{kPa}$ and tested undrained by increasing the axial stress with the total mean stress p held constant. It is observed that the change of pore pressure can be approximated by $\Delta u = \Delta g^2/300$.

The test results are given in Table 11.3 for equal increments of q. An undrained stress path defining a constant volume section of the state boundary surface or the wet side of critical is shown in Fig. 11.15.

Table 11.3

q' (kPa)	p (kPa)	u (kPa)	p' (kPa)
0	600	0	600
50	600	8	592
100	600	33	567
150	600	75	525
200	600	133	467
250	600	208	392
300	600	300	300

Figure 11.15

FURTHER READING

Atkinson, J. H. and P. L. Bransby (1978) *The Mechanics of Soils*, McGraw-Hill, London.
Muir Wood, D. M. (1991) *Soil Behaviour and Critical State Soil Mechanics*, Cambridge University Press, Cambridge.
Schofield, A. N. and C. P. Wroth (1968) *Critical State Soil Mechanics*, McGraw-Hill, London.

TWELVE
CAM CLAY

12.1 INTRODUCTION

Figure 11.7 shows a simple state boundary surface for soil; to develop a simple theoretical model for the stress–strain behaviour of soil this could be taken to be a yield surface. Yield curves are the lines of intersection of elastic walls with the yield surface as shown in Fig. 11.9 and these could be taken to be plastic potentials. We could then use the ideas of yielding, hardening and normality set out in Chapter 3 to derive a set of constitutive equations for soil. All that is required is a mathematical expression for the shape of the boundary surface.

Suitable equations for the state boundary surface could be obtained by fitting expressions to laboratory test data, by purely theoretical consideration of the mechanics of granular materials or by a combination of these. A very simple and neat theoretical equation was obtained by research workers in the University of Cambridge during the 1960s and this will be described here. Over the years many others have tried to improve on the original Cambridge equation and while some have succeeded in obtaining better agreement with experimental observations the simplicity and elegance of the original is inevitably lost. What I am going to do in this chapter is to describe the original simple theoretical model to get across the basic techniques involved in constructing constitutive equations for soil. Anyone seriously interested in applying these techniques in practice will need to study the more complex, and more realistic, soil models.

12.2 BASIC FEATURES OF THE CAM CLAY MODELS

The Cambridge theories are known under the umbrella term of Cam clay. The first model described by Schofield and Wroth (1968) is known as original Cam clay and a second model described by Roscoe and Burland (1968) is known as modified Cam clay. All the theories within the Cam clay family are basically similar. Soil is taken to be frictional with logarithmic compression. The state boundary surface is taken as a yield surface and as a plastic potential surface, and hardening is related to the plastic volumetric strains. The principle differences between the various members of the Cam clay family are in the precise equations used to describe the yield curves. For example, in original Cam clay they are logarithmic spirals while in modified Cam clay they are ellipses.

The term Cam clay was coined by the Cambridge research workers because the river in Cambridge is called the Cam. Do not misunderstand this. You cannot go to Cambridge and dig up any Cam clay; it is simply the name of a theoretical model or a set of equations. The status of Cam clay is like the status of elasticity. You cannot find any elasticity anywhere; what you can find is steel or copper which behave in a way very like the theory of elasticity, at least over small strains. In the same way you cannot find any Cam clay; what you can find are reconstituted (and some intact) soils that behave in a way very like the theoretical model called Cam clay.

12.3 STATE BOUNDARY SURFACE FOR ORDINARY CAM CLAY

The basic equation for the state boundary surface for ordinary Cam clay is

$$\frac{q'}{Mp'} + \left(\frac{\lambda}{\lambda - \kappa}\right) \ln p' - \left(\frac{\Gamma - v}{\lambda - \kappa}\right) = 1 \tag{12.1}$$

This defines the state boundary surface shown in Fig. 12.1. The surface meets the $v : p'$ plane along the isotropic normal compression line where $q' = 0$ and $v = N - \lambda \ln p'$ and hence, substituting into Eq. (12.1),

$$N - \Gamma = \lambda - \kappa \tag{12.2}$$

The curves shown in Fig. 12.1 are constant volume sections and undrained stress paths. The equation for an undrained stress path can be obtained from Eq. (12.1) with $v = \Gamma - \lambda \ln p'_c$, where p'_c is the stress at the intersection of the constant volume section and the critical state line and is

$$\frac{q'}{Mp'} + \left(\frac{\lambda}{\lambda - \kappa}\right) \ln \left(\frac{p'}{p_c}\right) = 1 \tag{12.3}$$

A yield curve is the intersection of an elastic wall given by $v = v_\kappa - \kappa \ln p'$ with the state boundary surface. At the critical state line the specific volume is v_c and the mean stress is p'_c, as shown in Fig. 12.2(b) where $v_c = v_\kappa - \kappa \ln p'_c = \Gamma - \lambda \ln p'_c$. Eliminating v and v_κ, the equation for the yield curve shown in Fig. 12.2(a) is

$$\frac{q'}{Mp'} + \ln \left(\frac{p'}{p_c}\right) = 1 \tag{12.4}$$

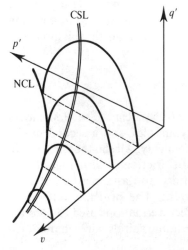

Figure 12.1 State boundary surface for ordinary Cam clay.

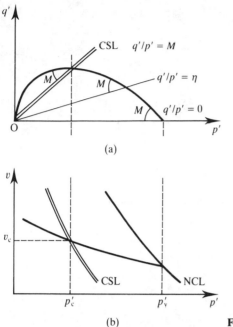

(a)

(b)

Figure 12.2 Yield curve for ordinary Cam clay.

Note that the equations of the constant volume section, or undrained stress path, and the yield curve are different except for the special case of a soil with $\kappa = 0$. From Eq. (12.4), with $q' = 0$, the yield stress p'_y is related to the critical state stress p'_c on the same yield curve by

$$\frac{p'_y}{p'_c} = \exp(1) = 2.72 \tag{12.5}$$

Differentiating Eq. (12.4) we get

$$\frac{dq'}{dp'} = \frac{q'}{p'} - M \tag{12.6}$$

which is simply another way of writing an equation for a yield curve. Equation (12.6) shows that the logarithmic spiral curve has the very simple property that the gradient dq'/dp' is related to the gradient q'/p' of the radius from the origin.

12.4 CALCULATION OF PLASTIC STRAINS

The yield curve is taken to be a plastic potential so that the vector of plastic strain increment $\delta\varepsilon^p$ is normal to the curve, as shown in Fig. 12.3. If two lines are orthogonal the product of their

Figure 12.3 Plastic potential and plastic strains for Cam clay.

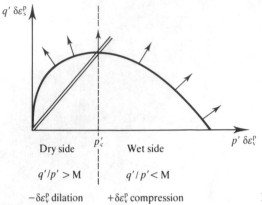

Dry side p'_c Wet side

$q'/p' > M$ $q'/p' < M$

$-\delta\varepsilon_v^p$ dilation $+\delta\varepsilon_v^p$ compression

Figure 12.4 Vectors of plastic strain for Cam clay.

gradients is -1, so

$$\frac{dq'}{dp'} \cdot \frac{d\varepsilon_s^p}{d\varepsilon_v^p} = -1 \tag{12.7}$$

and, from Eq. (12.6), the plastic strain increments are given by

$$\frac{\delta\varepsilon_v^p}{\delta\varepsilon_s^p} = M - \frac{q'}{p'} \tag{12.8}$$

At the critical state when $q'/p' = M$ we have $\delta\varepsilon_v^p = 0$. On the wet side $q'/p' < M$ and so $\delta\varepsilon_v^p$ is positive (i.e. compressive), while on the dry side $q'/p' > M$ and so $\delta\varepsilon_v^p$ is negative (i.e. dilative), as shown in Fig. 12.4.

Notice that Eq. (12.8) is almost the same as Eq. (10.12); the only difference is that Eq. (10.12) gives total strains while Eq. (12.8) gives the plastic strains. Equation (10.12) was obtained by analogy with the work done by friction and dilation and the derivation was for peak states on the dry side. The similarity between Eqs (10.12) and (12.8) demonstrates that the basis of ordinary Cam clay is an equivalent work equation, but now extended to the wet side as well as the dry side. A more rigorous derivation of ordinary Cam clay from work principles was given by Schofield and Wroth (1968).

12.5 YIELDING AND HARDENING

As the state moves on the state boundary surface from one yield curve to another there will be yielding and hardening (or softening if the state is on the dry side) and, in Cam clay, the change of the yield stress is related to the plastic volume change. Figure 12.5 shows an increment of loading A → B on the wet side of critical, and the state moves from one yield curve to a larger one with an increase in yield stress and a reduction in volume. The increment of loading C → D on the dry side is associated with a decrease in yield stress and an increase in volume. Equation 12.1) can be rewritten as

$$v = \Gamma + \lambda - \kappa - \lambda \ln p' - \frac{(\lambda - \kappa)q'}{Mp'} \tag{12.9}$$

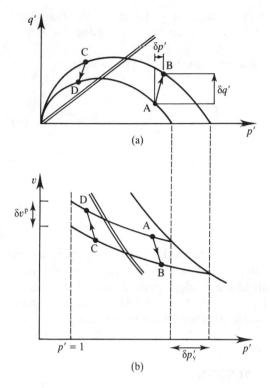

(a)

(b)

Figure 12.5 Hardening and softening for Cam clay.

Differentiating, dividing by v and noting that $\delta\varepsilon_v = -\delta v/v$, we have

$$\delta\varepsilon_v = \left(\frac{\lambda - \kappa}{vp'M}\right)\delta q' + \left[\frac{\lambda}{vp'} - \frac{(\lambda - \kappa)\eta'}{vp'M}\right]\delta p' \tag{12.10}$$

If we now subtract the elastic volumetric strains given by Eq. (11.3), the plastic volumetric strains are

$$\delta\varepsilon_v^p = \left(\frac{\lambda - \kappa}{vp'M}\right)[\delta q' + (M - \eta')\delta p'] \tag{12.11}$$

and, from Eq. (12.8), the plastic shear strains are

$$\delta\varepsilon_s^p = \left(\frac{\lambda - \kappa}{vp'M}\right)\left(\frac{\delta q'}{M - \eta'} + \delta p'\right) \tag{12.12}$$

12.6 COMPLETE CONSTITUTIVE EQUATIONS FOR ORDINARY CAM CLAY

The complete constitutive equations for Cam clays are obtained simply by adding the elastic strains given by Eqs (11.3) and (11.4) to the plastic strains given by Eqs (12.11) and (12.12) to obtain

$$\delta\varepsilon_s = \frac{1}{vp'}\left\{\left[\frac{\lambda - \kappa}{M(M - \eta')} + \frac{g}{3}\right]\delta q' + \left[\frac{\lambda - \kappa}{M}\right]\delta p'\right\} \tag{12.13}$$

$$\delta\varepsilon_v = \frac{1}{vp'}\left\{\left[\frac{\lambda - \kappa}{M}\right]\delta q' + \left[\frac{\lambda - \kappa}{M}(M - \eta') + \kappa\right]\delta p'\right\} \tag{12.14}$$

These apply for states that are on the state boundary surface; for states inside the boundary surface the elastic strains given by Eqs (11.3) and (11.4) can be recovered by putting $\lambda = \kappa$ into Eqs (12.13) and (12.14).

Equations (12.13) and (12.14) are constitutive equations like Eq. (3.18) and components of the compliance matrix are

$$C_{11} = \frac{1}{vp'}\left[\frac{\lambda - \kappa}{M(M - \eta')} + \frac{g}{3}\right] \tag{12.15}$$

$$C_{22} = \frac{1}{vp'}\left[\frac{\lambda - \kappa}{M}(M - \eta') + \kappa\right] \tag{12.16}$$

$$C_{12} = C_{21} = \frac{1}{vp'}\left[\frac{\lambda - \kappa}{M}\right] \tag{12.17}$$

These demonstrate that in Cam clay the basic compliances contain the intrinsic soil parameters M, λ, κ and g and the current state given by v, p' and $\eta' = q'/p'$. Thus, in Cam clay, the behaviour is non-linear since, in general, v, p' and q' change during a loading path. Notice that towards failure at the critical state when $\eta' \to M$ we have $C_{11} \to \infty$ and $C_{22} \to 0$. Thus, near ultimate failure, shear strains become very large while volumetric strains become very small.

12.7 APPLICATIONS OF CAM CLAY IN DESIGN

Although Eqs (12.13) and (12.14) are a complete set of constitutive equations for soil there is still quite a lot of further analysis required before they can be used for detailed design calculations. For example, they are written in terms of shearing and volumetric effects, but for calculations they need to be rewritten in terms of the normal and shear stresses and strains on horizontal and vertical planes in the ground and possibly in three dimensions.

Ordinary Cam clay has the advantage that with yield curves as logarithmic spirals the algebra is relatively simple. Although it describes the main features of soil behaviour qualitatively there are a number of detailed aspects where it is not so good. Another model, modified Cam clay, is based on yield curves that are ellipses; this is described in detail by Muir Wood (1991).

The Cam clay equations can be implemented in finite element and similar numerical analyses as described by Britto and Gunn (1987). Be warned though: these analyses are quite complex and difficult to do properly. If you are interested in making use of these advanced techniques you are advised to start by working with people who have previous experience.

12.8 SUMMARY

1. Cam clay is a theoretical model for soil behaviour: it includes strength and stress–strain behaviour within a single, relatively simple model.
2. Cam clay combines the theories of critical state soil mechanics and the idea of a state boundary surface with the theories of plasticity, including yielding, hardening and plastic flow.
3. There are different versions of Cam clay depending on the precise equation for the state boundary surface.

WORKED EXAMPLES

Example 12.1: Calculation of strains for overconsolidated Cam clay   A soil has the parameters $M = 0.98$, $\lambda = 0.20$ and $\kappa = g = 0.05$ and its behaviour can be represented by the Cam clay model. A sample is isotropically compressed in a stress path triaxial cell to $p' = 300$ kPa and swelled to $p'_0 = 200$ kPa where the specific volume is $v_0 = 2.13$. It is then subjected to a drained test in which $\delta q' = \delta p' = 10$ kPa.

The strains are given by Eqs (12.13) and (12.14) with $\lambda = \kappa$, since the state of the overconsolidated sample is inside the state boundary surface. Hence,

$$\delta\varepsilon_s = \frac{g}{3vp'}\,\delta q' = \frac{0.05 \times 10 \times 100}{2.13 \times 200 \times 3} = 0.04\%$$

$$\delta\varepsilon_v = \frac{\kappa}{vp'}\,\delta p' = \frac{0.05 \times 10 \times 100}{2.13 \times 200} = 0.11\%$$

Example 12.2: Calculation of strains for normally consolidated Cam clay   A second sample of the soil in Example 12.1 was isotropically compressed to $p'_0 = 200$ kPa where the specific volume was $v_0 = 2.19$. It was then subjected to a drained test in which $\delta q' = \delta p' = 10$ kPa.

The strains are given by Eqs (12.13) and (12.14), with the initial state $p' = 200$ kPa, $v = 2.19$ and $\eta' = 0$ corresponding to isotropic compression. The compliances given by Eqs (12.15) to (12.17) are

$$C_{11} = \frac{1}{200 \times 2.19}\left(\frac{0.15}{0.98^2} + \frac{0.05}{3}\right) = 0.39 \times 10^{-3}\ \text{m}^2/\text{kN}$$

$$C_{22} = \frac{1}{200 \times 2.19}(0.15 + 0.05) = 0.46 \times 10^{-3}\ \text{m}^2/\text{kN}$$

$$C_{12} = \frac{1}{200 \times 2.19}\left(\frac{0.15}{0.98}\right) = 0.35 \times 10^{-3}\ \text{m}^2/\text{kN}$$

and, hence,

$$\delta\varepsilon_s = (C_{11}\,\delta q' - C_{12}\,\delta p') \times 100 = 0.74\%$$

$$\delta\varepsilon_v = (C_{12}\,\delta q' + C_{22}\,\delta p') \times 100 = 0.81\%$$

REFERENCES

Britto, A. M. and M. J. Gunn (1987) *Critical State Soil Mechanics via Finite Elements*, Ellis Horwood, Chichester.

Muir Wood, D. M. (1991) *Soil Behaviour and Critical State Soil Mechanics*, Cambridge University Press, Cambridge.

Roscoe, K. H. and J. B. Burland (1968) 'On the generalised stress–strain behaviour of "wet" clay', in *Engineering Plasticity*, J. Heyman and F. A. Leckie (eds), Cambridge University Press, Cambridge.

Schofield, A. N. and C. P. Wroth (1968) *Critical State Soil Mechanics*, McGraw-Hill, London.

FURTHER READING

Atkinson, J. H. and P. L. Bransby (1978) *The Mechanics of Soils*, McGraw-Hill, London.

Muir Wood, D. M. (1991) *Soil Behaviour and Critical State Soil Mechanics*, Cambridge University Press, Cambridge.

Schofield, A. N. and C. P. Wroth (1968) *Critical State Soil Mechanics*, McGraw-Hill, London.

13.1 INTRODUCTION

Stiffness relates increments of stress and increments of strain. A knowledge of soil stiffness is required to calculate ground movements and to obtain solutions to problems of soil–structure interaction, such as loads on retaining walls. Often simple analyses are carried out assuming that soil is linear and elastic and solutions for foundations will be considered in Chapter 22. However, it is recognized that soil strains are often significantly inelastic and more complicated elasto-plastic models such as Cam clay (see Chapter 12) have been developed to model the stress–strain behaviour of soil.

The stress–strain behaviour of soil is actually more complex than that given by the simple Cam clay model, particularly at small strains and for states inside the state boundary surface where, in the simple theory, the strains are elastic. A detailed treatment of soil stiffness is beyond the scope of this book. What I am going to do in this chapter is simply describe the essential features of the stress–strain behaviour of soil as an introduction to further studies.

13.2 CAM CLAY AND SOIL STIFFNESS

In Chapter 12 the basic ideas of the classical theories of elasticity and plasticity were combined with the basic soil mechanics theories of friction and logarithmic compression into a general model known as Cam clay. A set of non-linear constitutive equations was obtained in terms of the intrinsic soil parameters λ, M, Γ, κ and g, together with parameters describing the current state and the loading history.

The basic equations for Cam clay for states on the state boundary surface (Eqs 12.13 and 12.14) contain elastic and plastic components of straining, while for states inside the state boundary surface the basic equations (Eqs 11.3 and 11.4) contain only elastic strains. It turns out that the ordinary Cam clay equations are reasonably good for states on the state boundary surface (the modified Cam clay equations are a little better), but the basic Cam clay theories are rather poor for states inside the state boundary surface where the behaviour is taken to be elastic and recoverable.

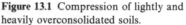

Figure 13.1 Compression of lightly and heavily overconsolidated soils.

The consequences of this for geotechnical design are illustrated in Fig. 13.1. This shows two soils subjected to exactly the same loading paths A → B and C → D. The soil which starts from A is lightly overconsolidated; it yields at Y when the state reaches the state boundary surface and then it moves along Y → B on the state boundary surface with elastic and plastic strains. The soil which starts from C is heavily overconsolidated, the state does not reach the state boundary surface and the strains are taken to be elastic throughout the loading path C → D. The stress–strain curves are shown in Fig. 13.1(b) and these correspond to the volume changes shown in Fig. 13.1(c).

For lightly overconsolidated soils following the path A → B in Fig. 13.1, the greater proportion of the strains occur along Y → B as the state moves on the state boundary surface and only a small proportion occurs along A → Y, where the soil is inside the boundary surface. For these soils we can use the Cam clay theories to calculate ground movements since the significant errors which occur in the calculations of the elastic strains along A → Y will be relatively small compared with the total strains for the whole path A → B. For heavily overconsolidated soils, on the other hand, the state remains inside the state boundary surface for the whole path C → D and the errors in the strains calculated using the Cam clay theories will be relatively large.

13.3 STIFFNESS–STRAIN RELATIONSHIPS FOR SOIL

From Eqs (3.13) to (3.17) a general set of constitutive equations can be written as

$$\begin{Bmatrix} \delta q' \\ \delta p' \end{Bmatrix} = \begin{bmatrix} 3G' & J' \\ J' & K' \end{bmatrix} \begin{Bmatrix} \delta \varepsilon_s \\ \delta \varepsilon_v \end{Bmatrix} \tag{13.1}$$

where G' is the shear modulus, K' is the bulk modulus and J' are moduli that couple shear and

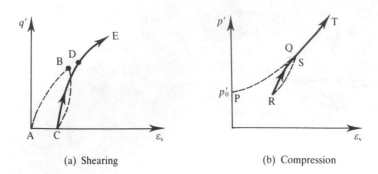

(a) Shearing (b) Compression

Figure 13.2 Shearing and compression of soils.

volumetric effects. For undrained loading for which $\delta\varepsilon_v = 0$, we have

$$\frac{dq'}{d\varepsilon_s} = 3G' \tag{13.2}$$

$$\frac{dp'}{d\varepsilon_s} = J' \tag{13.3}$$

and, for isotropic compression for which $\delta\varepsilon_s = 0$, we have

$$\frac{dp'}{d\varepsilon_v} = K' \tag{13.4}$$

$$\frac{dq'}{d\varepsilon_v} = J' \tag{13.5}$$

Notice that for undrained loading Eq. (13.2) also defines the undrained shear modulus G_u and hence

$$G_u = G' \tag{13.6}$$

Figure 13.2 shows the general characteristics of shearing and compression stress–strain curves for undrained shearing and isotropic compression tests with stages of loading, unloading and reloading. In Fig. 13.2(a) the gradient of the curve is the shear modulus $3G'$ and in Fig. 13.2(b) the gradient is the bulk modulus K'; we could obtain similar curves and evaluate J'_1 and J'_2 by plotting $\delta q'$ against $\delta\varepsilon_v$ and $\delta p'$ against $\delta\varepsilon_s$. In Fig. 13.2 the soil had been unloaded from B and from Q and so the initial states C and R are inside the state boundary surface and the soil yields at D and S.

In Fig. 13.2 the stress–strain lines CDE and RST look non-linear, but it is difficult to see exactly how the soil is behaving, especially for small increments at the start of the reloading. The principal features of the stress–strain curves can be seen more clearly if the stiffness is plotted against the strain. Figure 13.3(a) shows a typical shear modulus–shear strain curve for overconsolidated soil and Fig. 13.3(b) is a typical bulk modulus–volumetric strain curve: note that the strains are plotted to a logarithmic scale in each case.

The stiffness–strain curves for shear and bulk moduli shown in Fig. 13.3 are typical for soil. (Surprisingly, the general shape applies for normally consolidated soils as well as for lightly and heavily overconsolidated soil and the consequences of this will be considered later.) The curves for shear and bulk modulus are basically similar, except at strains in excess of 1 per cent or so. At large strains the shear modulus continues to decrease and becomes zero at ultimate failure, while the bulk modulus starts to increase, as shown in Fig. 13.2.

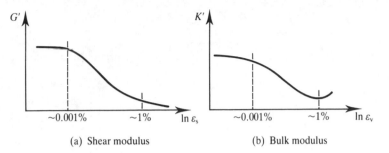

(a) Shear modulus (b) Bulk modulus

Figure 13.3 Characteristic stiffness–strain curves for soil.

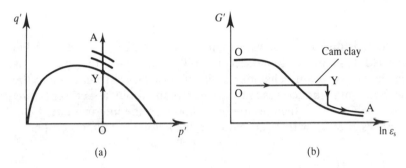

(a) (b)

Figure 13.4 Characteristic stress–strain behaviour for soil observed in laboratory tests and given by the Cam clay theories.

This stress–strain behaviour is significantly different from that given by the simple Cam clay theory described in Chapter 12. Figure 13.4 illustrates characteristic stress–strain behaviour observed in laboratory tests and given by Cam clay. For the drained constant p' loading path O → Y → A in Fig. 13.4(a), the state reaches the state boundary surface at Y and travels on the boundary surface along Y → A. For Cam clay the behaviour is taken to be elastic along O → Y and, since p' and v remain constant for the particular loading path considered, the shear modulus $G' = vp'/g$ remains constant, as shown in Fig. 13.4(b). When the state reaches the state boundary surface at Y, yield occurs and the stiffness drops sharply to the value given by the full Cam clay expression in Eq. (12.13). Figure 13.4(b) indicates that after yield the behaviour observed in laboratory tests will be very like that given by the Cam clay theories (with suitable values for the soil parameters), but before yield the observed stiffness–strain behaviour is very different in character from that given by Cam clay.

The principal features of soil stiffness are illustrated in Fig. 13.5. There are three regions,

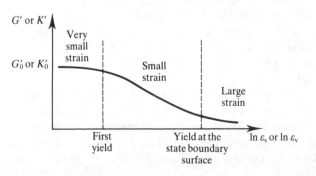

Figure 13.5 Characteristic ranges of soil stiffness.

(a) Deformations of a retaining wall   (b) Settlement of a foundation

Figure 13.6 Strains in the ground near typical geotechnical structures.

as indicated, where the behaviour is different. For very small strains, smaller than some value corresponding to the first yield (usually of the order of 0.001 per cent), the stiffness is approximately constant and the stress–strain behaviour is linear. For large strains, where the state has reached the state boundary surface (usually greater than about 1 per cent), the behaviour is elasto-plastic and the Cam clay theories are quite good. In the intermediate, small strain, range the stiffness changes rapidly with strain and the behaviour is highly non-linear.

13.4 STRAINS IN THE GROUND

In most geotechnical structures that are designed to restrict ground movements, such as foundations and retaining walls, the strains in the ground are usually quite small. Figure 13.6 illustrates a stiff retaining wall and a foundation. The outward movement of the top of the wall and the settlement of the foundation are both 10 mm and these would be acceptable displacements in many designs. The mean shear strains in the ground near the wall and the volumetric strains below the foundation are 0.1 per cent. In practice there will be local strains greater than these, especially near the edge of the foundation, and the strains will decay to zero far from the structures. This means that in the ground soil stiffness will vary continuously with position and with loading throughout most of the range illustrated in Fig. 13.5.

13.5 MEASUREMENT OF SOIL STIFFNESS IN LABORATORY TESTS

The best method for investigating soil stiffness and evaluating stiffness parameters is to conduct stress path triaxial tests in the laboratory using one of the hydraulic triaxial cells described in Sec. 7.9. This apparatus permits tests to be carried out in which the initial state and the loading path can be controlled. The principal problem arises in the measurement of the small and very small strains required to investigate the whole of the characteristic stiffness strain curves shown in Fig. 13.5. To examine the whole of the stiffness–strain curve it is necessary to measure strains less than 0.001 per cent; if the length of the sample is about 100 mm you will need to measure displacements smaller than 0.001 mm or 1 μm.

The problem is not so much with the resolution and accuracy of the dial gauges, displacement transducers and volume gauges used to measure axial and volumetric strains in triaxial tests as with the errors that occur due to compliance, or movement, in the apparatus. (Do not forget the distinction between accuracy and resolution discussed in Chapter 7.) Figure 13.7 illustrates a conventional triaxial test; the axial displacement ΔL is measured using a displacement

(a) Externally   (b) Local gauges   **Figure 13.7** Methods of strains in triaxial tests.

transducer or dial gauge mounted on the loading ram and the volume change is measured from the volume of pore water entering or leaving the sample through the drainage leads. Errors arise due to (1) axial displacements at the ends of the sample, (2) displacements where the loading ram joins the top platen, (3) movements in the load cell and (4) movements in the cell.

The errors that can arise due to the compliances illustrated in Fig. 13.7(a) can be very significant and can easily swamp the required measurements of small strains. In conventional triaxial tests the measured axial strains are unreliable at strains smaller than about 0.1 per cent irrespective of the resolution and accuracy of the transducer or dial gauge. If a hydraulic triaxial cell is used and if very careful measurements are made of the displacements in the apparatus, it is possible to obtain reliable measurements of axial and volumetric strains smaller than 0.01 per cent. One way to improve the accuracy of measurements of strain in triaxial tests is to use a gauge inside the cell mounted directly on the sample, as shown in Fig. 13.7(b). Using these kinds of instruments strains smaller than 0.001 per cent can be measured reliably.

It is very difficult to measure the stiffness of soil at very small strains (i.e. less than about 0.001 per cent) in triaxial tests by direct observations of strains. The simplest method is to calculate the shear modulus from the velocity of dynamic waves. The very small strain shear modulus G_0' is given by

$$G_0' = \frac{\gamma V_s^2}{g} \tag{13.7}$$

where V_s is the velocity of shear waves through the sample, γ is the unit weight of the soil and $g = 9.81$ m/s². Shear waves can be generated and their velocity measured directly using shear elements set into the top and bottom platens or from resonant frequencies in torsional shearing. The equipment and techniques for making these measurements are rather specialized and if you need to determine G_0' you will need help; it is enough now to know that the techniques are available.

Note that in these dynamic tests the rates of loading are very large and saturated soil will be undrained. This does not matter for measurement of shear modulus since, for shearing alone, $G' = G_u$. The undrained bulk modulus of saturated soil is theoretically infinite (since $\delta \varepsilon_v = 0$ for undrained loading) and so we cannot determine the small strain bulk modulus K_0' of saturated soil from the velocity of compression waves. (The velocity of compression waves through saturated soil is approximately the same as the velocity of sound in water, 1500 m/s).

Figure 13.8 summarizes the principal features of the application and measurement of soil

Figure 13.8 Characteristic ranges of stiffness in the field and in laboratory tests.

stiffness over a wide range of strain. In the field, strains in the ground near retaining walls and below foundations are relatively small and are usually less than 1 per cent, except in small regions near the edges of foundations. Stiffness cannot be measured reliably in ordinary triaxial tests at strains less than about 0.1 per cent unless special procedures are followed, so the ordinary triaxial test is not much good for measuring soil stiffness in the range of practical interest. Stiffness at small strains can be measured reliably using local gauges attached to the sample and the shear modulus at very small strain G'_0 can be obtained from measurements of shear wave velocity.

13.6 STIFFNESS OF SOIL AT SMALL AND VERY SMALL STRAINS

At large strains (i.e. greater than about 1 per cent) the state of lightly or heavily overconsolidated soil will have reached the state boundary surface and the stiffness parameters in Eq. (13.1) depend on the current state (v, p' and η') as given by Eqs (12.15) to (12.17). For states inside the state boundary surface, at small and very small strains, soil stiffness is highly non-linear, but we might expect that the stiffness at a particular strain will also depend on the current state and on the history.

(a) Stiffness at Very Small Strain

In dynamic tests used to measure G'_0, samples are vibrated at a constant state at strains less than about 0.001 per cent. The damping is negligible and at very small strains soil is linear and elastic. (If a typical value for G'_0 is 100 MPa then a strain $\delta\varepsilon_s = 0.001$ per cent corresponds to an increment of stress $\delta q'$ of only 3 kPa.)

The general relationship between G'_0 and the current state is of the form

$$\frac{G'_0}{p'_r} = A \left(\frac{p'}{p'_r}\right)^n R_p^m \tag{13.8}$$

where p'_r is a reference pressure included to make Eq. (13.8) dimensionless and A, m and n depend on the nature of the soil (Viggiani, 1992). Notice that in Eq. (13.8) the value of G'_0 is related to p' and R_p without the specific volume or voids ratio. This is possible because v, p' and R_p are not independent as discussed in Sec. 8.3 and so v is included in the parameters p' and R_p. Alternatively, G'_0 could be related to v and R_p. The value of the exponent n is generally in the

Figure 13.9 Typical variation of very small strain stiffness of soil with stress and overconsolidation.

range 0.5 to 1.0 and typical values for m are in the range 0.2 to 0.3. Equation (13.8) can be rewritten as

$$\ln\left(\frac{G'_0}{p'_r}\right) = \ln A + m \ln R_p + n \ln\left(\frac{p'}{p'_r}\right) \tag{13.9}$$

Plotting data from a set of tests carried out at different values of p' and R_p, as shown in Fig. 13.9, provides a convenient method for evaluating the parameters A, m and n.

(b) Stiffness at Small Strain

The general relationships between shear modulus G' and strain, state and history for small strains in the range 0.001 to 1 per cent are illustrated in Fig. 13.10 and the same general relationships hold for the other stiffness parameters. The value of G'/p' depends on strain (because

(a)

(b)

Figure 13.10 Typical variation of small strain stiffness of soil with strain, stress and overconsolidation.

of the non-linearity) and on $\ln R_p$ and, at a particular strain,

$$\frac{G'}{p'} = AR_p^m \tag{13.10}$$

where $A = G'_{nc}/p'$ is the stiffness of normally consolidated soil at the same strain. (Notice that Eq. (13.8) reduces to Eq. (13.10) when $n = 1$.) When soil is normally consolidated its state is on the state boundary surface so values for G'_{nc} are given by Eq. (12.15). Values for m depend on the nature of the soil and on the strain. A number of other factors, such as a rest period at constant stress and a change in the direction of the stress path between successive loading stages, also effect soil stiffness, but the rate of loading has virtually no effect provided that the soil is either fully drained or fully undrained.

13.7 NUMERICAL MODELLING OF SOIL STIFFNESS

Equations (13.8) and (13.10) are convenient expressions relating the shear modulus to the current state and to the stress history and there will be similar expressions for the bulk modulus K'. However, to be of practical use for design, soil behaviour must be represented by mathematical expressions similar to those developed for Cam clay in Chapter 12, although these are likely to be more complex to take account of the non-linear behaviour for states inside the boundary surface. One possibility is to regard soil behaviour inside the state boundary surface as essentially elastic, but non-linear, and to use curve-fitting techniques to obtain an empirical expression relating shear modulus G' and bulk modulus K' to strain. This is the approach followed by Duncan and Chang (1970) and by Jardine *et al.* (1991). This method requires complex laboratory tests in which the stress paths mimic the *in situ* paths and numerical analyses that should stop and restart at each change in the direction of a stress path. An alternative approach is to regard soil behaviour in the small strain region as inelastic, with yielding and hardening with moving yield surfaces inside the state boundary surface. One approach is to adapt the Cam clay models by including additional yield surfaces (e.g. Mroz, Norris and Zienkiewicz, 1979; Atkinson and Stallebrass, 1991). In these models the parameters remain the fundamental parameters required by Cam clay together with additional parameters that describe the relative sizes of the additional yield surfaces.

At small strains in the region 0.001 to 1 per cent the general relationships between shear modulus, strain and stress shown in Fig. 13.10 are similar for normally consolidated and overconsolidated soils. Furthermore, unloading and reloading loops, like those illustrated in Fig. 13.2, result in substantial irrecoverable strains. These observations indicate that the basic rules governing stiffness of overconsolidated soils at small strains are similar to those for normally consolidated soil which, as we have seen, are essentially elasto-plastic and not purely elastic as assumed in the Cam clay theories.

All this is really quite advanced and any further discussion of developments in theories for soil stiffness at small strain is clearly beyond the scope of this book.

13.8 SUMMARY

1. The stress–strain behaviour of soil is highly non-linear over the whole range of loading except at very small strains less than about 0.001 per cent. There are three ranges of behaviour:

(a) very small strain (usually less than 0.001 per cent),
(b) small strain,
(c) large strains (for states on the state boundary surface).

2. For states on the state boundary surface the strains are relatively large and can be modelled reasonably using Cam clay or a similar elasto-plastic model.

3. For very small strains the stress–strain behaviour is approximately linear and the shear modulus is given by

$$\frac{G_0'}{p_r'} = A \left(\frac{p'}{p_r'} \right)^n R_p^m \tag{13.8}$$

where A, m and n depend on the nature of the soil.

4. For small strains the soil is highly non-linear: at a particular strain the shear modulus is given by

$$\frac{G'}{p'} = A R_p^m \tag{13.10}$$

where A and m depend both on the nature of the soil and on the strain.

REFERENCES

Atkinson, J. H. and S. E. Stallebrass (1991) 'A model for recent history and non-linearity in the stress–strain behaviour of overconsolidated soil', *Proceedings of 7th IACMAG '91, Cairns*, pp. 555–560.

Duncan, J. M. and C. Y. Chang (1970) 'Non-linear analysis of stress and strain in soils', *ASCE, J. of the Soil Mechanics and Foundation Engng Div.*, **96**, SM5, 1629–1653.

Jardine, R. J., D. M. Potts, H. D. St. John and D. W. Hight (1991) 'Some applications of a non-linear ground model', *Proceedings 10th European Conference on Soil Mechanics and Foundation Engineering, Florence*, Vol. 1, pp. 223–228.

Mroz, Z., V. A. Norris and O. C. Zienkiewicz (1979) 'Application of an anisotropic hardening model in the analysis of elasto-plastic deformaton of soils', *Geotechnique*, **29**, 1, 1–34.

Viggiani, G. (1992) 'Small strain stiffness of fine grained soils', PhD thesis, City University.

FURTHER READING

Atkinson, J. H. and G. Sallfors (1991) 'Experimental determination of stress–strain–time characteristics in laboratory and *in situ* tests', *Proceedings of 10th ECSMFE, Florence*, Vol. 3, pp. 915–956.

Wroth, C. P. and G. T. Houlsby (1985) 'Soil mechanics—property characterisation and analysis procedures', *Proceedings of 11th ICSMFE, San Fancisco*, pp. 1–55.

FOURTEEN

CONSOLIDATION

14.1 BASIC MECHANISM OF CONSOLIDATION

In Sec. 6.8 we saw that, in general, any undrained loading or unloading will create excess pore pressures $\bar{u}$ in the region of the loading. These excess pore pressures may be positive or negative with respect to the steady state pore pressures u_0 and give rise to hydraulic gradients that cause seepage flow. These seepage flows lead to volume changes that, in turn, are associated with the changes of effective stress as the excess pore pressures dissipate. As the excess pore pressures diminish the hydraulic gradients and rates of flow also diminish, so that the volume changes continue at a reducing rate. After a long time the seepage and volume changes will stop when the excess pore pressures and hydraulic gradients become zero and the pore pressures reach their steady state values.

The coupling of seepage due to hydraulic gradients with compression or swelling due to the resulting seepage flow and changes of effective stress is known as consolidation, and this process accounts for settlement of foundations with time, progressive softening of soil in excavations and other similar effects. In order to calculate the rate at which excess pore pressures reduce it is necessary to develop a simple theory for consolidation.

A general theory for three-dimensional consolidation is quite complicated and here I will consider a simpler theory for one-dimensional consolidation in which all seepage flow and soil strains are vertical and there is no radial seepage or strain. This is relevant to conditions in an oedometer test (see Sec. 7.6), as shown in Fig. 14.1(a), and in the ground below a wide foundation on a relatively thin layer of soil, as shown in Fig. 14.1(b). In both cases the seepage of water from within the body of the soil is vertical and upwards towards a surface drainage layer where the steady state pore pressure is always u_0 and the excess pore pressure is always zero.

14.2 THEORY FOR ONE-DIMENSIONAL CONSOLIDATION

Figure 14.2 shows an element in a consolidating soil. (Here all dimensions increase positively downwards to avoid difficulties with signs.) In a time interval δt the thickness changes by δh.

(a) Oedometer test

(b) Wide foundation

Figure 14.1 One-dimensional consolidation.

Figure 14.2 One-dimensional consolidation.

The flow of water through the element is one-dimensional and the rates of flow in through the top and out through the bottom are q and $q + \delta q$ respectively. From the definition of the coefficient of compressibility m_v given by Eq. (8.9),

$$\delta h = -m_v \, \delta z \, \delta \sigma' \tag{14.1}$$

The theory requires that m_v remains constant and so it is valid only for relatively small increments of stress. Since the soil grains are incompressible an equation of continuity relates the change of volume of the element to the change of flow through it:

$$A \, \delta h = -\delta q \, \delta t \tag{14.2}$$

Combining Eqs (14.1) and (14.2) and in the limit noting that q and σ' are both functions of z and t,

$$\frac{\partial q}{\partial z} = Am_v \frac{\partial \sigma'}{\partial t} \tag{14.3}$$

The rate of seepage flow is given by Darcy's law as

$$V = \frac{q}{A} = ki \tag{14.4}$$

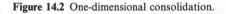

where V is the seepage velocity and the hydraulic gradient i is

$$i = -\frac{1}{\gamma_w}\frac{\delta \bar{u}}{\delta z} \tag{14.5}$$

From Eqs (14.4) and (14.5) and in the limit,

$$\frac{\partial q}{\partial z} = -\frac{Ak}{\gamma_w}\frac{\partial}{\partial z}\left(\frac{\partial \bar{u}}{\partial z}\right) = -\frac{Ak}{\gamma_w}\frac{\partial^2 \bar{u}}{\partial z^2} \tag{14.6}$$

and, from Eqs (14.3) and (14.6),

$$\frac{k}{m_v\gamma_w}\frac{\partial^2 \bar{u}}{\partial z^2} = -\frac{\partial \sigma'}{\partial t} \tag{14.7}$$

The effective stress is given by $\sigma' = \sigma - (u_0 + \bar{u})$ and, noting that u_0 remains constant,

$$\frac{\partial \sigma'}{\partial t} = \frac{\partial \sigma}{\partial t} - \frac{\partial \bar{u}}{\partial t} \tag{14.8}$$

The simple and common case is where consolidation takes place after an increment of undrained loading or unloading so that the total stress remains constant during the consolidation. Then, from Eqs (14.7) and (14.8) with $\partial\sigma/\partial t = 0$,

$$c_v\frac{\partial^2 \bar{u}}{\partial z^2} = \frac{\partial \bar{u}}{\partial t} \tag{14.9}$$

where

$$c_v = \frac{k}{m_v\gamma_w} \tag{14.10}$$

The parameter c_v is known as the coefficient of consolidation and has the units of square metres per year. Values of c_v depend on both the permeability k and on the compressibility m_v, both of which vary greatly for different soils.

Equation (14.9) is the basic equation for one-dimensional consolidation. Solutions will give the variations of excess pore pressure $\bar{u}$ with depth z and with time t. Note that consolidation theory deals with excess pore pressure $\bar{u}$ and not with absolute pore pressures.

14.3 ISOCHRONES

Solutions to Eq. (14.9) can be represented graphically by plotting the variation of $\bar{u}$ with depth at given times. The resulting family of curves are called isochrones (from the Greek and meaning equal time). A simple way to visualize isochrones is to imagine a set of standpipes inserted into the consolidating soil below a rapidly constructed embankment as shown in Fig. 14.3(a).

Before construction water rises in the standpipes to the steady state water table in the drain at the surface where the pore pressures are u_0. Undrained construction of the embankment adds a total stress $\Delta\sigma$ at the surface, which gives rise to initial excess pore pressures $\bar{u}_i = \Delta\sigma$ throughout the soil. The initial excess pore pressures registered by the standpipes are uniform with depth and water rises to the same height in all the pipes, as shown by the broken (initial) line in Fig. 14.3(a). The corresponding isochrone for $t = 0$ is shown in Fig. 14.3(b). (Notice that because $\gamma \approx 2\gamma_w$ the standpoints must project well above the maximum height of the embankment.)

At a time shortly after construction excess pore pressure at the top of the soil near the drain will have reduced to zero and excess pore pressures will have reduced elsewhere, so the variation of the levels of water in the standpipes is similar to that shown by the curved broken line. This

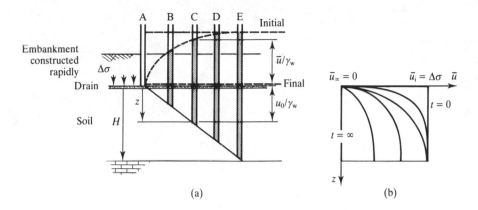

Figure 14.3 Isochrones for one-dimensional consolidation.

broken line gives the shape of the isochrone at a particular time. After a very long time all the excess pore pressures have dissipated and the levels of water in the standpipes are again at the steady state groundwater table; the isochrone for $t - \infty$ is the final broken line.

Figure 14.3(b) shows a set of isochrones for the one-dimensional consolidation illustrated in Fig. 14.3(a) plotted as $\bar{u}$ against depth z. Each isochrone corresponds to a particular time: for $t = 0$, $\bar{u}_i = \Delta\sigma$ at all depths and at $t = \infty$, $\bar{u}_\infty = 0$.

14.4 PROPERTIES OF ISOCHRONES

Isochrones must satisfy the one-dimensional consolidation equation together with the drainage boundary conditions, and these requirements impose conditions on the geometry and properties of isochrones. Consolidation, with dissipation to a drain at the surface, as shown in Fig. 14.3, starts near the surface and progresses down through the soil. At relatively small times, such as t_n in Fig. 14.4, consolidation is limited to the upper levels only and below a depth n the excess pore pressures have not fallen. At large times, such as t_m, consolidation is occurring throughout the layer. There is a critical time t_c when excess pore pressures at the base first start to dissipate; the isochrone for t_c is shown in Fig. 14.4(a). Figure 14.4(b) illustrates the dissipation of excess pore pressure at the different depths indicated in Fig. 14.4(a). Near the surface, at a depth z_1, the excess pore pressures dissipate very rapidly but near the base, at a depth z_3, the excess pore pressures remain at $\bar{u}_i$ until the critical time t_c.

The gradient of an isochrone is related to the hydraulic gradient by

$$\frac{\partial \bar{u}}{\partial z} = -\gamma_w i \tag{14.11}$$

and from Darcy's law the seepage velocity is

$$V = -\frac{k}{\gamma_w} \frac{\partial \bar{u}}{\partial z} \tag{14.12}$$

By inspection of the isochrones in Fig. 14.4(a) the gradients of the isochrones, and hence the seepage velocities, increase towards the surface. At the base of an isochrone there is no seepage flow, either because it represents the limit of consolidation for t_n or because of the impermeable boundary for t_m, and so the isochrones must be vertical at the base, as shown in Fig. 14.4(a).

(a)

(b)

Figure 14.4 Dissipation of excess pore pressure during consolidation.

Figure 14.5 Area between two isochrones.

Since soil grains and water are incompressible the velocity of the upward seepage at any level must equal the rate of settlement at that level and

$$\frac{\partial \rho}{\partial t} = \frac{k}{\gamma_w} \frac{\partial \bar{u}}{\partial z} \tag{14.13}$$

The movement of isochrones represents changes of excess pore pressure and changes of effective stress. Figure 14.5 shows isochrones for t_1 and t_2. From Eq. (14.1) the change of thickness δh of the thin slice δz is given by $\delta h = -m_v \, \delta z \, \delta \sigma'$. If the total stress remains constant, $\delta \sigma' = -\delta \bar{u}$ and

$$\delta h = m_v \, \delta z \, \delta \bar{u} \tag{14.14}$$

where $\delta z \, \delta \bar{u}$ is the shaded area in Fig. 14.5. Summing the changes of thickness for all thin slices in the depth z, the change of surface settlement between the times t_1 and t_2 is given by

$$\delta\rho = m_v \times \text{area OAB} \tag{14.15}$$

Hence the settlement of a consolidating layer in a given time is given by m_v times the area swept by the isochrone during the time interval.

14.5 SOLUTION FOR ONE-DIMENSIONAL CONSOLIDATION BY PARABOLIC ISOCHRONES

Simple and reasonably accurate solutions for the rate of settlement for one-dimensional consolidation can be obtained by assuming that the general shapes of the isochrones in Fig. 14.4(a) can be approximated by parabolas. It is necessary to treat the cases $t < t_c$ and $t > t_c$ separately; the ideas behind each analysis are the same but the algebra differs slightly.

(a) $t = t_n < t_c$

Figure 14.6(a) shows an isochrone for time t_n; the slope is vertical at N and no consolidation has occurred below a depth n. From Eq. (14.15) (noting that the area below a parabola is $\frac{1}{3} \times$ base $\times$ height), the surface settlement is given by

$$\Delta\rho_t = m_v \times \text{area AEN} = \tfrac{1}{3} m_v n \, \Delta\sigma \tag{14.16}$$

Differentiating Eq. (14.16) and noting that m_v and $\Delta\sigma$ are assumed to be constants during consolidation, the rate of settlement is given by

$$\frac{\mathrm{d}\rho_t}{\mathrm{d}t} = \frac{1}{3} m_v \, \Delta\sigma \frac{\mathrm{d}n}{\mathrm{d}t} \tag{14.17}$$

The rate of surface settlement is also related to the gradient of the isochrone at Λ. From Eq. (14.13) and noting that from the geometry of a parabola the gradient at A is $2\Delta\sigma/n$, we have

$$\frac{\mathrm{d}\rho_t}{\mathrm{d}t} = \frac{k}{\gamma_w} \frac{2\Delta\sigma}{n} \tag{14.18}$$

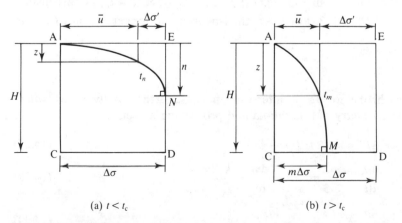

(a) $t < t_c$ (b) $t > t_c$

Figure 14.6 Geometry of parabolic isochrones.

Hence, equating the rates of surface settlement from Eqs (14.17) and (14.18),

$$n\frac{dn}{dt} = 6\frac{k}{m_v \gamma_w} = 6c_v \tag{14.19}$$

and, integrating with the boundary condition $n = 0$ at $t = 0$,

$$n = \sqrt{12c_v t} \tag{14.20}$$

Equation (14.20) gives the rate at which the effects of consolidation progress into the soil from the drainage boundary; no dissipation of excess pore pressure will occur at depths greater than n. Using Eq. (14.20) and the geometry of a parabola it is possible to calculate the excess pore pressure at any depth and at any time $t < t_c$.

In practice, the most important thing to calculate is the surface settlement $\Delta\rho_t$ after a time $t < t_c$; and this is found by substituting for n into Eq. (14.16), giving

$$\Delta\rho_t = \tfrac{1}{3}m_v \, \Delta\sigma\sqrt{12c_v t} \tag{14.21}$$

The final surface settlement $\Delta\rho_\infty$ will occur after a long time when all excess pore pressures have dissipated and $\Delta\sigma' = \Delta\sigma$. Hence, from Eq. (14.1),

$$\Delta\rho_\infty = m_v H \, \Delta\sigma \tag{14.22}$$

Combining Eqs (14.21) and (14.22),

$$\frac{\Delta\rho_t}{\Delta\rho_\infty} = \frac{2}{\sqrt{3}}\sqrt{\frac{c_v t}{H^2}} \tag{14.23}$$

Equation (14.23) may be written in terms of a dimensionless degree of consolidation U_t and a dimensionless time factor T_v given by

$$U_t = \frac{\Delta\rho_t}{\Delta\rho_\infty} \tag{14.24}$$

$$T_v = \frac{c_v t}{H^2} \tag{14.25}$$

and the general solution becomes

$$U_t = \frac{2}{\sqrt{3}}\sqrt{T_v} \tag{14.26}$$

This solution is valid until the point N in Fig. 14.6(a) reaches D when $t = t_c$; at this instant $n = H = \sqrt{12c_v t}$ so $T_v = \tfrac{1}{12}$ and $U_t = 0.33$. For $t > t_c$ the isochrone no longer touches ED and a new analysis is required.

(b) $t = t_m > t_c$

Figure 14.6(b) shows an isochrone for t_m; it intersects the base orthogonally at M where $\bar{u} = m \, \Delta\sigma$. Making use of the geometry of a parabola and proceeding as before,

$$\Delta\rho_t = m_v \, \Delta\sigma \, H(1 - \tfrac{2}{3}m) \tag{14.27}$$

$$\frac{d\rho_t}{dt} = -\frac{2}{3}m_v \, \Delta\sigma \, H\frac{dm}{dt} = \frac{k}{\gamma_w}\frac{2m \, \Delta\sigma}{H} \tag{14.28}$$

$$m\frac{dm}{dt} = -\frac{3c_v}{H^2} = -\frac{1}{t}3T_v \tag{14.29}$$

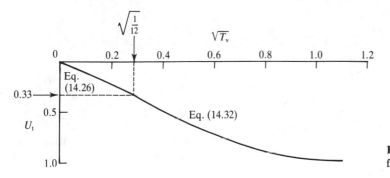

Figure 14.7 Solution for consolidation from parabolic isochrones.

Integrating Eq. (14.29) between the limits $m = 1$ and $T_v = \frac{1}{12}$ at $t = t_c$ and $m = 0$ at $t = \infty$, we have

$$m = \exp(\tfrac{1}{4} - 3T_v) \tag{14.30}$$

Equation (14.30), together with the geometry of a parabola, may be used to calculate the excess pore pressure at any depth and at any time $t > t_o$. Proceeding as before, the surface settlement and the degree of consolidation are given by

$$\Delta\rho_t = m_v H\,\Delta\sigma[1 - \tfrac{2}{3}\exp(\tfrac{1}{4} - 3T_v)] \tag{14.31}$$

$$U_t = 1 - \tfrac{2}{3}\exp(\tfrac{1}{4} - 3T_v) \tag{14.32}$$

The complete solution for one-dimensional consolidation with parabolic isochrones consists of Eq. (14.26) for $T_v < \frac{1}{12}$ and Eq. (14.32) for $T_v > \frac{1}{12}$, as shown in Fig. 14.7. For most practical purposes consolidation can be taken to be completed at $T_v = 1$. Excess pore pressures can be found from the geometry of the parabolic isochrones shown in Fig. 14.6 with values for n and m calculated from Eqs (14.20) and (14.30) respectively.

Notice that in all the examples discussed so far drainage has been one-way to the upper surface and the base was impermeable, as illustrated in Fig. 14.8(a). Often in practice and in laboratory tests the drainage is two-way to drains at the top and bottom, as illustrated in Fig. 14.8(b). In this case the soil consolidates as two symmetric halves, each with one-way drainage, and the rate of consolidation is governed by H^2. We can avoid ambiguity by redefining H as the maximum drainage path; thus H in Eq. (14.25) is the longest direct path taken by a drop of water as it is squeezed from the soil.

(a) One-way drainage   (b) Two-way drainage

Figure 14.8 Boundary drainage conditions for one-dimensional consolidation.

Figure 14.9 Solutions for one-dimensional consolidation.

14.6 OTHER CONSOLIDATION SOLUTIONS

The solutions obtained using parabolic isochrones are simple and illustrative but are restricted to the case of one-dimensional consolidation where the initial excess pore pressure $\bar{u}_i$ is the same everywhere. Other solutions are available for other cases.

The one-dimensional consolidation equation can be solved analytically and the solution is in the form of a Fourier series (Taylor, 1948). The degree of consolidation is given by

$$U_t = 1 - \sum_{m=0}^{\infty} \frac{2}{M^2} \exp(-M^2 T_v) \tag{14.33}$$

where $M = \frac{1}{2}\pi(2m + 1)$. For values of U_t not greater than about 0.6, Eq. (14.33) can be approximated to

$$U_t = \frac{2}{\sqrt{\pi}} \sqrt{T_v} \tag{14.34}$$

which is close to Eq. (14.26) which is the solution using parabolic isochrones for small times.

The solutions will be slightly different if the initial excess pore pressures are not everywhere the same. The two common cases are where the initial excess pore pressures increase or decrease linearly with depth. Relationships between U_t and $\sqrt{T_v}$ for three cases of initial excess pore pressure are shown in Fig. 14.9.

14.7 DETERMINATION OF c_v FROM OEDOMETER TESTS

The results of a single stage of consolidation of a sample in an oedometer test may be used to estimate a value for the coefficient of consolidation of a soil. Since the time factor T_v is a function of c_v, we cannot immediately plot experimental results of U_t against T_v. However, if the test is continued until consolidation is complete, we may find the final settlement ρ_∞ and, hence, the degree of consolidation at any time, and thus plot U_t against time t. If the experimental U_t against t curve can be fitted to a theoretical U_t against T_v curve, a relationship between t and T_v may be obtained and c_v found from Eq. (14.25). Two alternative curve-fitting approximations are available.

(a) A $\sqrt{}$(Time) Method

This method makes use of the observation that settlement against $\sqrt{}$(time) curves have an initial portion that may be approximated by a straight line, and this straight line can be fitted to Eq.

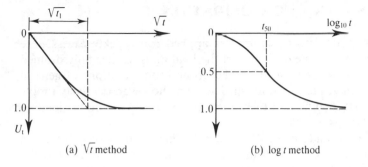

Figure 14.10 Determination of c_v from oedometer test results by curve fitting.

(a) $\sqrt{t}$ method

(b) $\log t$ method

(14.34). Figure 14.10(a) shows the results of a single stage of consolidation of a sample of clay in an oedometer test plotted as U_t against $\sqrt{t}$. The slope of the initially linear part of the curve is given by $\sqrt{t_1}$, as shown in Fig. 14.10(a). The experimental curve and the curve in Fig. 14.10(a) fit when $U_t = 1$ and $t = t_1$ in Eq. (14.25). Hence,

$$\sqrt{T_v} = \sqrt{\frac{C_v t_1}{H^2}} = \frac{\sqrt{3}}{2} \tag{14.35}$$

$$c_v = \frac{3H^2}{4t_1} \tag{14.36}$$

(b) A Log$_{10}$ (Time) Method

As an alternative, it is sometimes more convenient to fit the experimental and theoretical consolidation curves at $U_t = 0.5$, i.e. when half of the consolidation is complete. The value of T_v for $U_t = 0.5$ may be found from Eq. (14.33) and is $T_v = 0.196$. To estimate a value for t_{50}, the time for $U_t = 0.5$ during a single stage of consolidation in an oedometer test, it is convenient to plot U_t against $\log t$ as shown in Fig. 14.10(b). The value for t_{50} may be read directly from the experimental consolidation curve. Theoretical and experimental curves fit when

$$T_v = \frac{c_v t_{50}}{H^2} = 0.196 \tag{14.37}$$

$$c_v = 0.196 \left(\frac{H^2}{t_{50}} \right) \tag{14.38}$$

Note that U_t cannot be calculated until the final settlement $\Delta \rho_\infty$ has been found. Ideally, settlement–time curves would approach horizontal asymptotes as illustrated in Fig. 14.10 and it would not be difficult to estimate a value for $\Delta \rho_\infty$. For most experimental settlement–time curves, however, these horizontal asymptotes are not clearly defined and, moreover, there is often an initial settlement which is observed immediately after the loading increment has been applied. For most practical cases it is necessary to estimate a value for $\Delta \rho_\infty$ by means of special constructions. A construction for estimating $\Delta \rho_\infty$ from a plot of $\Delta \rho_t$ against $\sqrt{t}$ was proposed by Taylor and a construction for estimating $\Delta \rho_\infty$ from a plot of $\Delta \rho_t$ against $\log_{10} t$ was proposed by Casagrande; both constructions are described by Taylor (1948).

14.8 CONTINUOUS LOADING AND CONSOLIDATION

If the loading in a test which is supposed to be drained is applied too quickly excess pore pressures will occur but there will also be some drainage, so the loading is neither fully drained nor fully undrained. This is, of course, what happens in the ground, but solutions of general problems of coupled loading and drainage are very difficult. There are, however, relatively simple solutions for coupled one-dimensional loading and these form the basis of continuous loading consolidation tests (Atkinson and Davison, 1990).

Figure 14.11(a) shows a continuous loading one-dimensional compression test with a drain at the top and an impermeable boundary at the bottom. At a particular instant in the test the total stress is σ, the settlement is ρ and the pore pressures at the top and bottom of the sample are u_0 and u_b, so the excess pore pressure at the base is $\bar{u}_b = u_b - u_0$. The shaded area in Fig. 14.11(b) is $\sigma'H$, where σ' is the mean vertical effective stress and the isochrone is taken to be parabolic. Figure 14.11(c) shows the variations of total stress σ, settlement ρ and pore pressures u_0 and u_b, all of which must be measured during the test.

From Eqs (14.7) and (14.8) the basic equation for coupled loading and consolidation is

$$c_v \frac{\partial^2 \bar{u}}{\partial z^2} = \frac{\partial \bar{u}}{\partial t} - \frac{\partial \sigma}{\partial t} = -\frac{\partial \sigma'}{\partial t} \tag{14.39}$$

If the rate of loading is sufficiently slow so that $\bar{u}_b$ is relatively small compared with $\sigma - u_0$, then the mean effective stress can be approximated by $\sigma' = \sigma - u_0$. From the definition of the

Figure 14.11 Behaviour of soil during continuous loading consolidation tests.

coefficient of compressibility m_v given by Eq. (8.9),

$$m_v = -\frac{1}{H}\frac{dH}{d\sigma'}$$

(14.40)

If the isochrone is a parabola then the excess pore pressure at any depth z is given by

$$\bar{u}_z = \bar{u}_b\left(\frac{2z}{H} - \frac{z^2}{H^2}\right)$$

(14.41)

Differentiating twice,

$$\frac{d^2\bar{u}}{dz^2} = -\frac{2\bar{u}_b}{H^2}$$

(14.42)

and, substituting into Eq. (14.39),

$$c_v = \frac{H^2}{2\bar{u}_b}\frac{d\sigma'}{dt}$$

(14.43)

Then, from Eqs (14.40) and (14.43) together with Eq. (14.10),

$$k = \frac{\gamma_w H}{2\bar{u}_b}\frac{dH}{dt}$$

(14.44)

The compression consolidation and permeability parameters, m_v, c_v and k, can be evaluated from any one-dimensional continuous loading test in terms of the current values of simple thickness H and the excess pore pressure at the undrained face $\bar{u}_b$ and the gradients $d\sigma'/dH$, $d\sigma'/dt$ and dH/dt. In a test in which the sample dimensions and pore pressures are recorded at frequent intervals, values for the gradients may be determined by a numerical procedure and the values for the soil parameters calculated at equally frequent intervals.

14.9 SUMMARY

1. Consolidation occurs when excess pore pressures dissipate, usually at constant total stress. This results in compression or swelling as the effective stresses change.
2. The basic equation of one-dimensional consolidation is

$$c_v\frac{\partial^2\bar{u}}{\partial z^2} = \frac{\partial\bar{u}}{\partial t}$$

(14.9)

where the coefficient of consolidation is $c_v = k/m_v\gamma_w$, which has the units of square metres per year. Values of c_v can be determined from results of oedometer tests.
3. Solutions to Eq. (14.9) are represented by isochrones, which show the variation of excess pore pressure with time throughout the consolidating layer. Simple solutions for one-dimensional consolidation can be found, assuming that the isochrones are parabolas.
4. Standard solutions for consolidation settlements are given in terms of the degree of

consolidation and the time factor:

$$U_t = \frac{\Delta \rho_t}{\Delta \rho_\infty} \tag{14.24}$$

$$T_v = \frac{c_v t}{H^2} \tag{14.25}$$

Relationships between U_t and T_v depend on the distribution of the initial excess pore pressures and the drainage geometry.

WORKED EXAMPLES

Example 14.1: Interpretation of oedometer test results   The first two columns of Table 14.1 contain data from a single increment of an oedometer test in which the total vertical stress was raised from $\sigma = 90$ kPa to $\sigma = 300$ kPa. At $t = 0$ the sample was 20 mm thick and it was allowed to drain from the top and from the bottom.

Table 14.1

Time (min)	Settlement Δp_t (mm)	U_t	$\sqrt{t}$ (min$^{1/2}$)	log t
0	0	0	0	—
0.25	0.206	0.107	0.5	−0.602
1	0.414	0.216	1	0
2.25	0.624	0.325	1.5	0.352
4	0.829	0.432	2	0.602
9	1.233	0.642	3	0.954
16	1.497	0.780	4	1.204
25	1.685	0.878	5	1.398
36	1.807	0.941	6	1.556
49	1.872	0.975	7	1.690
24 h	1.920	1.000	—	—

For two-way drainage the drainage path is $H = 10$ mm. The degree of consolidation U_t is given by Eq. (14.24), taking the final settlement as $\Delta \rho_\infty = 1.920$ mm corresponding to $t = 24$ h.

(a) $\sqrt{t}$ method. Figure 14.12(a) shows U_t plotted against $\sqrt{t}$. Scaling from the diagram, $\sqrt{t_1} = 4.6$ and hence $t_1 = 21.2$ min. From Eq. (14.36),

$$c_v = \frac{3H^2}{4t_1} = \frac{3 \times (10 \times 10^{-3})^2}{4 \times 21.2} \times 60 \times 24 \times 365 = 1.9 \text{ m}^2/\text{year}$$

(b) Log t method. Figure 14.12(b) shows U_t plotted against log t. From the figure, log $t_{50} = 0.70$ and $t_{50} = 5.01$ min. From Eq. (14.38),

$$c_v = \frac{0.196H^2}{t_{50}} = \frac{0.196 \times (10 \times 10^{-3})^2}{5.01} \times 60 \times 24 \times 365 = 2.1 \text{ m}^2/\text{year}$$

The mean value for the coefficient of consolidation from the two methods is $c_v = 2$ m^2/year.

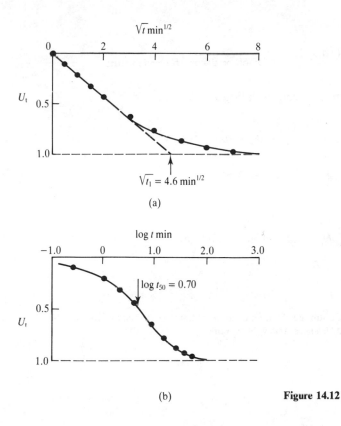

(a)

(b)   **Figure 14.12**

During the increment the vertical effective stress changes from $\sigma' = 90$ kPa at the start to $\sigma' = 300$ kPa at the end. The vertical strain is $\Delta\varepsilon_z = 1.920/20 = 0.096$ and from Eq. (8.9) the coefficient of compressibility is

$$m_v = \frac{\Delta\varepsilon_z}{\Delta\sigma'_z} = \frac{0.096}{300 - 90} = 4.6 \times 10^{-4}\ \text{m}^2/\text{kN}$$

From Eq. (14.10) the coefficient of permeability is given by

$$k = c_v m_v \gamma_w = \frac{2.0 \times 4.6 \times 10^{-4} \times 9.81}{60^2 \times 24 \times 365} = 2.9 \times 10^{-10}\ \text{m/s}$$

Example 14.2: Settlement of an oedometer sample   In a stage of an oedometer test the total stress was raised by 100 kPa. The sample was initially 20 mm thick and it was drained from both ends. The properties of the soil were $c_v = 2$ m²/year and $m_v = 5 \times 10^{-4}$ m²/kN.
 From Eq. (14.1) the final settlement, after consolidation is complete, was

$$\rho = m_v z\, \Delta\sigma'_z = 5 \times 10^{-4} \times 20 \times 100 = 1.0\ \text{mm}$$

(a) The time factor at which the settlement will be 0.25 mm (i.e. when $U_t = 0.25$) is given by Eq. (14.26):

$$T_v = \frac{3U_t^2}{4} = \frac{3 \times 0.25^2}{4} = 0.05$$

From Eq. (14.25), taking $H = 10$ mm for two-way drainage, the time when the settlement is 10 mm is

$$t = \frac{T_v H^2}{c_v} = \frac{0.05 \times (10 \times 10^{-3})^2}{2} \times 60 \times 24 \times 365 = 1.3 \text{ min}$$

(b) After 3 min the time factor and degree of consolidation are

$$T_v = \frac{c_v t}{H^2} = \frac{2 \times 3}{(10 \times 10^{-3})^2 \times 60 \times 24 \times 365} = 0.11$$

$$U_t = \frac{2}{\sqrt{3}} \sqrt{T_v} = \frac{2 \times \sqrt{0.11}}{\sqrt{3}} = 0.39$$

and the settlement is

$$\rho_t = \rho_\infty U_t = 1.0 \times 0.39 = 0.39 \text{ mm}$$

REFERENCES

Atkinson, J. H. and L. R. Davison (1990) 'Continuous loading oedometer tests', *Q. J. Engng Geol.*, **23**, 347–355.
Taylor, D. W. (1948) *Fundamentals of Soil Mechanics*, Wiley, New York.

CHAPTER FIFTEEN

AGEING AND STRUCTURE IN NATURAL SOILS

15.1 CHARACTERISTICS OF NATURAL SOILS

In Chapters 8 to 14 I described the basic mechanics of soils and in Chapters 17 to 22 these simple theories for soil behaviour will be used to investigate the performance of soil structures such as slopes, retaining walls and foundations. The behaviour described and the theories developed were largely idealizations for the behaviour of reconstituted soils, but natural soils differ from reconstituted soils in a number of important aspects.

Most natural soils are naturally deposited (from wind, water or ice) in changing depositional environments and so they are likely to have layers and lenses of different material. (Go and look carefully at freshly excavated slopes in soils and you will almost always be able to see layering; occasionally you can find deep beds of nearly uniform clay deposited in an unchanging environment, but these are rare.) Natural soils are then compressed and swelled one-dimensionally (i.e. with zero horizontal strain) by deposition and erosion, by weight of ice or by changing groundwater. They remain in the ground for very long periods of time (soil 10 000 years old is relatively very young) and they may experience physical and chemical changes. These changes are known collectively as ageing and include phenomena like cementing and weathering. Natural soils that contain one or more of these features are often called structured while reconstituted samples which have been completely disturbed and reconsolidated are sometimes called destructured.

It is very difficult to discover the true behaviour of natural soils. The obvious way is to recover undisturbed samples from the ground and test them in the laboratory but, unfortunately, the process of recovering the sample from the ground and installing it in the test apparatus will alter its state and its behaviour. There is no possibility of recovering and testing a truly undisturbed sample; the best we can do is to take and test an intact sample with the very minimum of disturbance. It is always understood that if the correct procedures for sampling and testing are followed the behaviour of an intact sample will be very close to the behaviour of the soil in the ground, but it is essential to follow the correct procedures.

This book deals with the basic, simple theories of soil mechanics relevant to reconstituted soils and a detailed discussion of all the effects and consequences of ageing and structure in natural soils is beyond its scope. It is, however, important to note these effects, which is the purpose of this chapter. The important thing is to consider the behaviour of intact samples of natural soils within the basic simple framework developed for reconstituted soils.

15.2 FORMATION OF NATURAL SOILS: ONE-DIMENSIONAL COMPRESSION AND SWELLING

The behaviour of soils during one-dimensional compression and swelling in laboratory tests was discussed in Sec. 8.4 and similar behaviour will occur during deposition and erosion of soil in the ground. Figure 15.1(a) illustrates a soil element below a ground level which rises and falls due to deposition and erosion and Fig. 15.1(b) shows the resulting changes of effective stress and water content. So far I have considered volume and volume changes in terms of the specific volume v or the voids ratio e, but in this chapter I shall consider water content w, as this is a commonly measured and often quoted parameter. Water content, specific volume and voids ratio are simply related (see Sec. 5.6) and $e = wG_s$. At points A and B the soil is normally consolidated and at C it is overconsolidated. Notice that although the vertical stresses at A and C are similar the water contents are very different. Figure 15.1(c) illustrates the changes of vertical and horizontal effective stresses during deposition and erosion. These can be related by a coefficient of earth pressure at rest, K_0, given as

$$K_0 = \frac{\sigma'_h}{\sigma'_z} \tag{15.1}$$

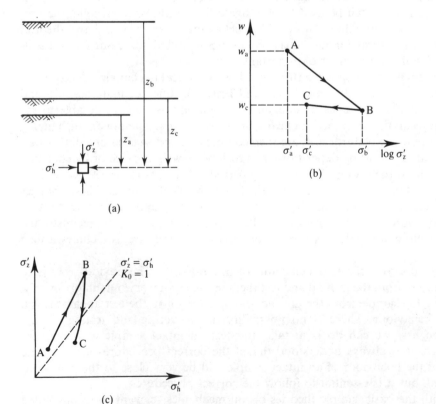

Figure 15.1 One-dimensional consolidation and swelling of soil in the ground due to deposition and erosion.

Figure 15.2 Variations of water content in the ground in normally consolidated and overconsolidated soils.

For normally consolidated and lightly overconsolidated soils $\sigma'_h < \sigma'_z$ and $K_0 < 1$, while for heavily overconsolidated soils $\sigma'_h > \sigma'_z$ and $K_0 > 1$. An approximation often used to estimate K_0 is

$$K_0 = K_{0nc}\sqrt{R_0} \tag{15.2}$$

where R_0 is the overconsolidation ratio and $K_{0nc} = 1 - \sin \phi'_c$ is the value of K_0 for normally consolidated soil.

In previous chapters I showed that many aspects of soil behaviour (but not the critical states) depend on the history of loading and unloading. This means that reconstituted samples should be compressed and swelled one-dimensionally in the triaxial apparatus before shearing and intact samples of natural soil should be reconsolidated to the estimated state in the ground.

The state of an element of soil in the ground depends on the current stresses (i.e. on the depth) and on the overconsolidation ratio (i.e. on the current depth and on the depth of erosion). Figure 15.2 illustrates the variations of water content with depth for a deposit which is lightly eroded (i.e. the depth of erosion z_e is small) or heavily eroded (i.e. the depth of erosion is large). For the lightly eroded soil the difference between the water contents at A and B is relatively large, while for the heavily eroded soil the difference between the water contents at C and D is much smaller and the water contents themselves are smaller. For the heavily eroded soil the smaller variation of water content with depth is a result of the very large maximum past stress.

Figure 15.3 Overconsolidation due to deposition and erosion.

(You should demonstrate this for yourself by calculating and plotting the variation of water content with depth for soil with $z_e = 1$ and 100 m, taking reasonable values for e_0, C_c and C_s in Eqs (8.10) and (8.11).)

15.3 AGEING

In the simple theories of soil mechanics plastic volume changes can only occur during loading on the state boundary surface. In Fig. 15.3 the state at A can move to B only by loading along the normal compression line: the irrecoverable plastic water content change δw^p is associated with a change of the yield stress σ'_{ya} to σ'_{yb} and a change in the apparent overconsolidation ratio from R_a to R_b.

Natural soils were deposited long ago: London Clay is about 60 million years old and even recent glacial soils are over 10 000 years old. Occasionally you may come across soils like Mississippi delta muds or the Fens in East Anglia which are only decades or centuries old, but these are very much the exception. As soils remain in the ground for very long periods, possibly without any loading or unloading due to deposition, erosion or groundwater changes, all kinds of things will happen to them. I will use the term ageing for all the processes, except loading, unloading and seepage of water, that occur in soils with time.

The most important of these ageing processes are creep, cementing, weathering, compaction and changes in the salinity of the pore water. These processes are equivalent to changes of the overconsolidation ratio at constant effective stress.

15.4 VIBRATION AND COMPACTION

If sand or gravel is vibrated at constant effective stress or compacted by impact or rolling it will compress and there will be irrecoverable plastic volume changes and changes in the overconsolidation ratio. Figure 15.4(a) shows the state path A → B corresponding to compaction. The yield point has increased from σ'_{ya} to σ'_{yb} with a corresponding increase in the overconsolidation ratio from R_a to R_b. (Notice that the overconsolidation ratio has increased because the yield stress has increased and not because the soil has been unloaded.) Figure 15.4(b) shows the corresponding state path normalized with respect to σ'_c; this assumes that both σ'_z and σ'_h remain constant and so the change of state is due to the increase of σ'_c as the volume decreases.

(a) (b)

Figure 15.4 Overconsolidation due to compaction.

15.5 CREEP

The effects of volume changes due to creep illustrated in Fig. 15.5(a) are similar to those due to compaction, except that volume changes due to compaction occur more or less instantaneously whereas those due to creep occur slowly and at a rate that diminishes with time. The basic constitutive equation for creep given in Sec. 3.7 is of the form

$$\delta w = C_\alpha \ln\left(\frac{t}{t_0}\right) \tag{15.3}$$

and so the water content decreases with the logarithm of time, as illustrated in Fig. 15.5(b). This influence of creep on the apparent overconsolidation ratio for soft clays was clearly demonstrated by Bjerrum (1967).

(a) (b)

Figure 15.5 Overconsolidation due to creep.

15.6 CEMENTING

During compaction or creep the current state of the soil changes, but during cementing and weathering both the current state and the state boundary surface may change. A detailed discussion of the effects of cementing and weathering is beyond the scope of this book and all I can do here is outline the basic features; for more detailed discussion there are papers by Leroueil and Vaughan (1990) and Coop and Atkinson (1993).

The principal mechanism of cementing in soils is by deposition of additional material, often calcium carbonate, from the groundwater. This has the dual effect of reducing the specific volume (because additional solid material appears) and shifting the state boundary surface. However, remember that the critical state corresponds to relatively large straining when the soil is essentially reconstituted and to reach these states the cementing material must fracture.

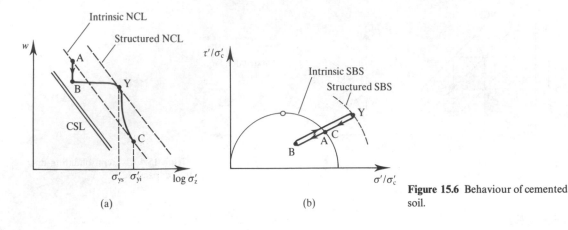

Figure 15.6 Behaviour of cemented soil.

This means that the critical states of cemented and uncemented material will be about the same and the principal influence of cementing will be on yielding and on the small strain stiffness.

The best way to examine cemented or structured soils is to carry out tests on both intact and reconstituted samples. Figure 15.6(a) illustrates a unique critical state line and an intrinsic normal compression line corresponding to reconstituted (i.e. destructured) soil. The path A → B represents a reduction in water content due to deposition of cementitious material at constant stress. The path B → Y → C represents compression of intact structured soil and part of this is outside the intrinsic normal compression line. The yield point Y lies on a structured normal compression line but, after yield, the state moves back towards the intrinsic line with increasing strain. There are now two yield points, σ'_{ys} associated with structured soil and σ'_{yi} associated with reconstituted material. Notice the relatively large compression from Y to C as the brittle cementing fractures.

There is, however, only one unique critical state line so values for the normalizing parameter σ'_c can be obtained unambiguously. (This is the principal reason for selecting σ'_c as the normalizing parameter rather than the equivalent pressure on the normal compression line σ'_e; see Sec. 9.5.) Figure 15.6(b) illustrates the state path A → B → Y → C normalized with respect to σ'_c. (The path is for loading with constant stress ratio.) Part of the loading path lies outside the intrinsic state boundary surface and the yield point Y lies on the structured state boundary surface. Notice that the states A and C lie at the same point and the state B is overconsolidated. The distance that the structured boundary surface lies outside the intrinsic boundary surface depends principally on the strength of the cementing.

15.7 WEATHERING

Weathering involves physical and chemical alteration of soils and rocks at essentially constant effective stress. This may very well alter the position of the intrinsic state boundary surface and the critical state line because the nature of the soil (i.e. its grading and mineralogy) changes. Weathering may also change the current state, usually by an increase in water content. There are no hard and fast rules for changes due to weathering and much depends on the nature of the weathering and on the initial soil and rock. The net effects of weathering are likely to be combinations of the other effects of ageing discussed above.

Figure 15.7 Compression of soil following changes of pore water salinity.

15.8 CHANGES IN PORE WATER SALINITY

The intrinsic properties of a soil depend principally on the nature of the grains but also on the chemistry of the pore water. Soil grains deposited through and compressed in saline water may have different intrinsic properties from the same soil grains deposited in fresh water. Figure 15.7 illustrates a soil normally compressed to A in saline water. If the salinity of the groundwater changes, perhaps because of changing sea level or uplift of the land, the intrinsic normal compression line may shift from the normal compression line corresponding to saline pore water NCL(S) to the normal consolidation line corresponding to fresh water NCL(F). On subsequent loading from A the soil will show relatively large compression as the state moves to B on the intrinsic normal compression line corresponding to fresh water NCL(F). Notice that the behaviour illustrated in Fig. 15.7 is similar to the behaviour of structured soil shown in Fig. 15.6.

15.9 SUMMARY

1. The state of a soil in the ground is determined primarily by the history of deposition and erosion, but it may be altered subsequently by the various processes of ageing.
2. The principal processes of ageing are compacting, creep, cementing, weathering and changes in the salinity of the pore water.
3. Ageing may change either the current state or the position of the state boundary surface.

REFERENCES

Bjerrum, L. (1967) 'Engineering geology of Norwegian marine clays', *Geotechnique*, **27**, 2, 81–118.
Coop, M. A. and J. H. Atkinson (1993) 'Mechanics of cemented carbonate sands', *Geotechnique*, [to be published].
Leroueil, S. and P. R. Vaughan, (1990) 'The general and congruent effects of structure in natural soils and weak rocks', *Geotechnique*, **50**, 3, 467–488.

CHAPTER SIXTEEN

GROUND INVESTIGATIONS

16.1 INTRODUCTION

Engineers designing structures and machines normally choose materials and specify their strength and stiffness and they often combine materials to make composites (e.g. steel and concrete in reinforced concrete). Similarly, highway engineers can specify the soils and rocks to be used in the construction of roads. Geotechnical engineers, on the other hand, cannot choose and must work with the materials in the ground. They must therefore determine what there is in the ground and the engineering properties of the ground, and this is the purpose of ground investigations.

The basic techniques of ground investigation are drilling, sampling and testing, *in situ* and in the laboratory, but these must be complemented by geological information and a sound appreciation of the relevant soil mechanics principles. Consequently, it is in the area of ground investigation that geology and engineering combine and where engineering geologists and geotechnical engineers cooperate.

Ground investigation is, of course, far too big a topic to be covered in one short chapter and all I will do here is outline the basic issues as a starting point for further study. The detailed techniques vary from country to country, and from region to region, and depend both on the local ground conditions, on historical precedents, on contractural procedures and on the available equipment and expertise. As with laboratory testing, procedures for ground investigations are covered by national standards and codes of practice; in the United Kingdom this is BS 5930:1981. You should look up the standards covering the region where you work in to see what they contain. Detailed descriptions of the current practices in the United Kingdom are given by Clayton, Simons and Matthews (1982).

16.2 OBJECTIVES OF GROUND INVESTIGATIONS

When you look at the face of a cliff or an excavation you see a section of the ground and when you look at a site you have to imagine what an excavation would reveal. A major part of a ground investigation is to construct a three-dimensional picture of the positions of all the

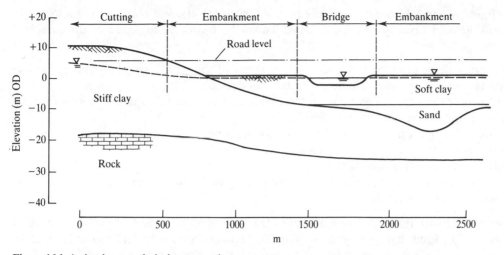

Figure 16.1 A simple geotechnical cross-section.

important soil and rock layers within the site that may be influenced by, or may influence, the proposed construction. Of equal importance is the necessity to sort out and identify the groundwater conditions. Note that in distinguishing the important soil and rock layers engineering classifications based on the nature and state of the soils (see Chapter 5) should be used rather than the geological classifications, which are based on age (see Chapter 4).

There is no simple answer to the problem of how many holes should be drilled and to what depths and how many tests should be carried out. Most of the standards and codes of practice make various recommendations, but really you should do enough investigation to satisfy everybody that safe and economical works can be designed and constructed.

Figure 16.1 illustrates a very simplified section along the centre-line of a road. (Notice that the horizontal and vertical scales are not the same.) The ground conditions revealed by drilling and other methods have been greatly idealized so that a number of characteristic layers have been identified and the boundaries between them drawn as smooth lines. The actual soils in the ground within any one layer are likely to be variable, horizontally and vertically, and their boundaries irregular. Something like Fig. 16.1 is about the best you can do with a reasonable investigation. Notice that Fig. 16.1 is a section along the centre-line of the road and to complete the investigation you should be able to draw cross-sections and sections on either side of the road.

The section shown in Fig. 16.1 is similar to that shown in Fig. 4.4(c) and I have already discussed the sequence of geological events and processes that formed this sequence of deposits. Certain features of the nature and state of the various layers can be estimated from consideration of their depositional environment and subsequent geological history. The grading and mineralogy of the soft clay and the stiff clay are the same (so they have the same nature), but their water contents are different (so they have different states); the soft clay is normally consolidated or lightly overconsolidated while the stiff clay is heavily overconsolidated.

For each of the principal strata in Fig. 16.1 you will need to determine representative parameters for strength, stiffness and water seepage flow (i.e. permeability). These will be selected from the results of laboratory and *in situ* tests. These parameters may be constant for a particular layer or they may vary with depth; generally we expect strength and stiffness to increase with depth. The parameters to be determined will be those that influence the design of the various structures in the works (i.e. the cutting in the stiff clay, the embankment on the soft clay and the bridge foundations, which will probably be piled, either to the stiff clay or the rock).

After any ground investigation you should know the following for each of the principal strata:

1. Its engineering description and classification in terms of the nature (grading and plasticity) and state (stress and specific volume or overconsolidation).
2. The positions of the boundaries between the different strata (i.e. you should be able to draw sections like that in Fig. 16.1 in any direction).
3. The geological environment when the soil was deposited and the history of subsequent deposition, erosion, weathering and ageing.
4. Descriptions of visible features of structure and fabric (e.g. layering, fissuring and jointing).
5. Representative values for the parameters for strength, stiffness and permeability relevant to the design and construction of the works.

You should also be sure that you know all about the groundwater. A very experienced ground engineer once said to me that he would not start an excavation until he knew exactly what he was digging into and what the groundwater conditions were; this is very good advice.

16.3 PLANNING AND DOING INVESTIGATIONS

You cannot really plan an entire ground investigation because you do not know what is there before you start and so you cannot select the best methods or decide how much to do. A ground investigation must, therefore, be carried out in stages: each stage can be planned with existing information and the knowledge gained from one stage will assist with planning the next. Currently in the United Kingdom a ground investigation is often let as a single contract with a specification and bill of quantities, which leads to major problems in planning the investigation and can often cause later difficulties.

There should be three principal stages in a ground investigation. (These are not rigid demarcations. There is often overlap between the stages; they need not be strictly sequential and one or other may have to be expanded later.)

(a) Desk Studies

This consists of study of all the information that you can find existing on paper. The major sources are topographical and geological maps and sections, geological reports and local authority records. Other sources include air photographs, historical archives and reports on earlier site investigations at the site or at nearby sites. Experienced geotechnical engineers and engineering geologists can often decipher the principal ground conditions from the desk study, so leading to well-planned later stages.

(b) Preliminary Investigations

Preliminary investigations are carried out at the site, rather than in the office, but they do not yet involve major expenditure on drilling, sampling and testing. The purposes are, firstly, to confirm or revise the findings of the desk study and, secondly, to add further information. This additional information will come from detailed engineering geological mapping, and this is best done by engineers and geologists working together or by experienced engineering geologists. Preliminary investigations may also involve some limited sub-surface exploration by trial pits, probing or exploratory drilling and geophysical sensing using seismic, electrical resistivity and other methods.

(c) Detailed Investigations

Detailed investigations consist of drilling, sampling and laboratory and *in situ* testing. They may also involve more detailed geological mapping, groundwater and chemical studies and other appropriate investigations necessary for the works. This is where the bulk of the expenditure is incurred and planning of the detailed investigations should set out to discover the required facts in the most efficient way. This will require some foreknowledge which can be gained from the desk study and preliminary investigations.

16.4 TEST PITTING, DRILLING AND SAMPLING

The standard method of ground investigation is excavation and sampling supplemented by *in situ* and laboratory testing. The excavations are usually done by drilling but also by opening test pits.

(a) Test Pitting

A test pit is an excavation that a geotechnical engineer or engineering geologist can enter to examine the soil profile *in situ*. Pits can be excavated by large drilling machines of the kind used for boring piles, by an excavator or by hand digging. Remember that any excavation in soil with vertical or steep sides is basically unstable and must be supported before anyone enters it.

(b) Drilling

Drill holes can be advanced into the ground using a number of different techniques; the principal kinds are illustrated in Fig. 16.2. Augers may be drilled to shallow depths by hand and large diameter augers can be drilled by machines used also for installation of bored piles (see Chapter 23). Wash boring is used in sands and gravels and rotary drilling is used mainly in rocks. Light percussion drilling is widely used in the United Kingdom and you can very often see the typical tripod rigs at work.

In some soils, particularly stiff clays and in rocks, boreholes will remain open unsupported, but in soft clays and particularly in coarse-grained soils the hole will need to be cased to maintain stability. Boreholes should normally be kept full of water, or bentonite mud, to prevent disturbance below the bottom of the hole.

(c) Sampling

Samples obtained from test pits or boreholes may be disturbed or intact. (Samples are often called disturbed or undisturbed but, as no soil sample is ever truly undisturbed, the word intact can be used for samples taken with minimum disturbance.) Disturbed samples are used principally for description and classification. Intact samples may be cut from the base or sides of test pits using saws or knives or taken in tubes pushed into the bottom of a borehole. There are many different tube samples; two designs used in the United Kingdom are shown in Fig. 16.3(a) and (b).

The tube sampler most often used in practice in the United Kingdom is the U100 illustrated in Fig. 16.3(a). The tube, nominally 100 mm in diameter, is screwed to a cutting shoe and a sampler head. The thickness of the wall of the cutting head is 6 to 7 mm, which is relatively large. A thin wall sample tube like that illustrated in Fig. 16.3(b) has a wall thickness of 1 to 2 mm and a cutting edge formed by machining. Both samplers are capable of taking samples in many

(a) Percussion drilling

(b) Flight auger drilling

(c) Wash boring

(d) Rotary drilling in rock

Figure 16.2 Methods for ground investigation drilling (schematic).

soft and stiff clays. Intact samples may be recovered by coring (see Fig. 16.3c), where a rotary drill cuts an annulus around the core sample. In the past this method was used exclusively for rocks but is now also used in stiff clays.

16.5 *IN SITU* TESTING

Laboratory tests to determine soil strength, stiffness and permeability are described in Chapter 7, but there are also a number of *in situ* tests. These can be grouped into probing tests, loading tests and permeability tests.

(a) Standard U100 tube   (b) Thin-wall tube   (c) Rotary coring

Figure 16.3 Methods of sampling in bore-holes (schematic).

(a) Probing Tests

In these tests a tool, usually cone-shaped, is hammered or pushed into the ground and the resistance to penetration recorded. This gives some measure of the strength and stiffness of the ground. In the standard penetration test (SPT) shown in Fig. 16.4(a) a solid cone or thick-wall tube is hammered, with a standardized blow, into the bottom of a borehole. The result is given as N, the number of blows to achieve a standard penetration; values increase from about 1 to more than 50 with increasing relative density or overconsolidation ratio.

In the static cone, or Dutch cone, penetration test shown in Fig. 16.4(b) the instrument is steadily pushed into the ground from the surface and the resistance recorded continuously. Most static cone penetrometers have a sleeve behind the cone which measures a frictional or shearing resistance. Some modern cones, known as piezocones, also measure pore pressures generated at

(a) Standard penetration test (SPT)   (b) Static (Dutch) cone testing   **Figure 16.4** Probing tests (schematic).

(a) Plate loading test

(b) Shear vane test

(c) Pressuremeter

Figure 16.5 *In situ* loading tests.

the tip or shoulder of the cone. Methods for interpretation of static cone tests were given by Meigh (1987). Many of these depend on empirical correlations between test observations and soil characteristics.

(b) Loading Tests

In these tests an instrument loads the soil in a controlled manner and stresses and deformations are observed. The ultimate load, when the deformations are large, is related to the strength of the soil and the load–deformation behaviour is related to soil stiffness. Plate loading tests illustrated in Fig. 16.5(a) may be carried out near the ground surface or at the bottom of a borehole and measurements are made of the load on the plate F and its settlement ρ. Simple analysis of plate tests are rather like the methods used for design of foundations, discussed in Chapter 22.

The shear vane test, illustrated in Fig. 16.5(b), is used to measure the undrained strength s_u. A vane with four blades is pushed into the ground from the surface or from the bottom of a borehole. The vane is rotated and the torque T measured. At the ultimate state the shear stress on the cylinder of soil containing the vane is given by

$$T = \tfrac{1}{2}\pi D^2 H\left(1 + \frac{1}{3}\frac{D}{H}\right)s_u \tag{16.1}$$

and a value for s_u can be calculated from the measured torque. If the rotation is continued for several revolutions the strength will drop to the residual (see Sec. 9.2).

Pressuremeter tests are illustrated in Fig. 16.5(c). A flexible cylinder is inflated and the pressures and volume changes measured. The best pressuremeters measure radial displacements directly (instead of volume changes) and some measure pore pressures as well. Pressuremeters may be installed in pre-drilled boreholes or self-boring devices drill themselves into the ground with less disturbance. Results of pressuremeter tests are used to calculate both soil strength, stiffness and the *in situ* horizontal stress σ_h. Methods for analysis of pressuremeter tests are described by Mair and Wood (1987).

16.6 STATES OF SOILS IN THE GROUND

As a soil is deposited in the ground, loaded by deposition of additional material and later unloaded by erosion, the state will be governed by the history of the stress changes together with any ageing effects, discussed in Chapter 15. If the history of deposition, erosion and groundwater changes can be determined from the geological history than it is often possible to make reasonable estimates of the likely engineering properties of soils in the ground.

Figure 16.6(b) illustrates compression from a water content near the liquid limit and swelling from a water content close to the plastic limit. The position of the critical state line is shown in Fig. 16.6(a), which is the same as Fig. 9.16(a). From Eq. (9.19), taking a typical value for $\phi_c' = 26°$, we have $\sigma_c' = 300$ kPa. The distance between the normal compression line and the critical state line in Fig. 16.6(b) depends on the nature of the soil, but typically for many clay soils $\sigma_e' \approx 2\sigma_c'$. Hence, for normally consolidated soil we have $\sigma_z' \approx 6$ kPa at the liquid limit and $\sigma_z' \approx 600$ kPa at the plastic limit; these correspond to depths in the ground of about 0.6 and 60 m respectively. Figure 16.6(c) illustrates typical profiles of water content with depth in the ground for a normally consolidated soil and a heavily overconsolidated soil. Notice that in a deposit of heavily overconsolidated soil the water content is close to the plastic limit, except near

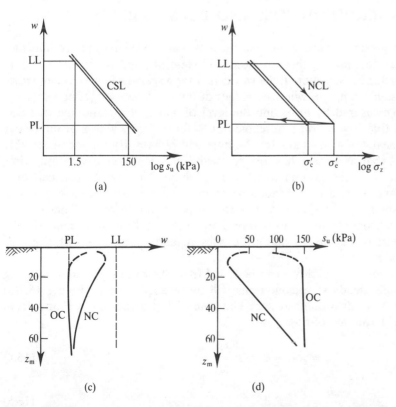

Figure 16.6 Water contents and undrained strengths in the ground in normally consolidated and overconsolidated soils.

the surface where the influence of rainwater in shallow cracks causes the water content to increase. In a deposit of normally consolidated or very lightly oveconsolidated soil the water content decreases with depth: near the surface the influence of vegetation and evaporation causes the water content to decrease.

The variations of critical state undrained strength with depth corresponding to the water contents in Fig. 16.6(c) are illustrated in Fig. 16.6(d). In the deposit of heavily overconsolidated soil the undrained strength s_u is approximately 150 kPa, corresponding to the water content close to the plastic limit, except near the surface where swelling causes the strength to decrease. In the deposit of normally consolidated soil the undrained strength increases linearly with depth, except near the surface where there is a higher strength in the crust due to the reduction in water content. The rate of increase of strength with depth, or with vertical effective stress, in a deposit of normally consolidated soil can be obtained from Eq. (9.19) as

$$\frac{s_u}{\sigma'_z} = \frac{\sigma'_c}{\sigma'_z} \tan \phi'_c \tag{16.2}$$

and taking typical values of $\phi'_c = 26°$ and $\sigma'_z/\sigma'_c = 2$ we have $s_u/\sigma'_z = 0.25$. An empirical relationship between s_u/σ'_z and PI was given by Skempton (1957) as

$$\frac{s_u}{\sigma'_z} = 0.11 + 0.0037\text{PI} \tag{16.3}$$

This is used widely to estimate the undrained strength of soft clays.

16.7 INVESTIGATING GROUNDWATER AND PERMEABILITY

Whatever else you do in a ground investigation you must be sure to define the groundwater conditions. This will include determining the current steady state pore pressures and the final steady state pore pressures after construction. If the works involve a seepage flow of water, either steady state or during consolidation, you will need values of the coefficient of permeability.

Pore pressures can be measured by observing the level of water in a standpipe (see Sec. 17.1) in a borehole. Notice that if you drill a borehole into saturated clay with a groundwater table, or phreatic surface, near the ground surface the hole will remain dry for a considerable time. The reason for this is that if the clay has low permeability it will take a very long time for sufficient water to flow from the ground to fill the borehole. This means you can only determine pore pressures, and groundwater conditions, from observation in boreholes in soils with relatively high permeability. For clays and soils with low permeability you will need to use special piezometers (i.e. instruments to measure pore pressures). In the final analysis the groundwater conditions must be reasonable and self-consistent and compatible with the soils and the regional hydrogeology.

Values for the coefficient of permeability k can be found from the results of *in situ* pumping tests. For coarse-grained soils steady state conditions will be reached quickly. Figure 16.7(a) illustrates steady state flow towards a pumped well. The potential at a radius r is P and, from Darcy's law (see Chapter 17), the rate of flow q is

$$q = Aki = 2\pi r P k \frac{dP}{dr} \tag{16.4}$$

or

$$\frac{dr}{r} = \frac{2\pi k}{q} P \, dP \tag{16.5}$$

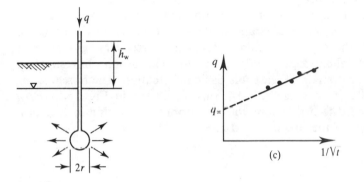

(a) Well pumping tests

(b) Piezometer tests

Figure 16.7 *In situ* permeability tests.

(Notice that the hydraulic gradient at the phreatic surface is strictly dP/ds, but dP/dr is a sufficiently good approximation.) Integrating Eq. (16.5) between P_1 at r_1 and P_2 at r_2 we have

$$\ln\left(\frac{r_2}{r_1}\right) = \frac{\pi k}{q}(P_2^2 - P_1^2) \tag{16.6}$$

Hence k can be obtained from observations of the pumping rate q and water levels in standpipes at a number of different radii.

For fine-grained soils steady state seepage will not be reached quickly and during a reasonable test period there will be simultaneous steady state flow and consolidation or swelling. Figure 16.7(b) illustrates a flow from a spherical cavity radius r with a constant excess pore pressure $\bar{u} = \gamma_w \bar{h}_w$. The rate of flow at any time t is given by

$$q = 4\pi r k \bar{h}_w \left(1 + \frac{r}{\sqrt{\pi c_s t}}\right) \tag{16.7}$$

where c_s is the coefficient of consolidation for spherical consolidation. (This is similar to c_v for one-dimensional flow, discussed in Chapter 14.) A condition of steady state flow would be reached after infinite time and, with $t = \infty$ in Eq. (16.7),

$$q_\infty = 4\pi r k \bar{h}_w \tag{16.8}$$

A value of q_∞ can be found by plotting q against $1/\sqrt{t}$, as shown in Fig. 16.7(c), and extrapolating. Hence a value for k can be obtained from Eq. (16.8). If the cavity is not spherical the term $4\pi r$ must be replaced by an intake factor F which depends on the geometry.

16.8 GROUND INVESTIGATION REPORTS

The findings of ground investigations are recorded in two different kinds of reports.

(a) Factual Reports

These simply describe the procedures and findings without comment or interpretation. The report will contain text describing what was done, how, where and by whom. It will summarize the factual findings of the desk study, the field investigations and the *in situ* and laboratory tests.

The basic information from the drilling and sampling operations is contained in borehole logs. (Similar logs contain information from test pits.) A typical borehole log is shown in Fig. 16.8; this is idealized and simplified to illustrate the principle features which should be recorded. The top panel gives the date, time, place, method of drilling and other basic information. The legend is a pictorial representation of the principal strata with a word description alongside. To the left are depths and levels. To the right are columns for sample recovery, groundwater observations and *in situ* tests. Borehole logs prepared by different ground investigation companies differ in detail but should contain at least this basic information. The borehole log in Fig. 16.8 is for a borehole drilled at chainage 2250 m on the section in Fig. 16.1. (How many more boreholes would you need to drill before you could draw the section in Fig. 16.1, given some idea of the basic geology of the site described in Sec. 4.6?)

(b) Interpretive Reports

An interpretive report will contain all the information in a factual report or it may refer to a separate factual report, but it will contain geological and engineering interpretations of the results of the investigations. An interpretive report should contain detailed engineering geological maps and sections giving a comprehensive three-dimensional picture of the engineering geology and hydrogeology of the site. For each of the principal soil and rock strata identified the interpretive report should give values for the parameters for strength, stiffness and permeability that will be used in the design. (These should relate to the requirements for the design of the individual structures in the scheme and the methods of analysis proposed.)

16.9 SUMMARY

1. In any geotechnical engineering activity investigations are required to determine the ground conditions. The objectives are to locate and identify all the principal soil and rock strata, estimate design values for their strengths and stiffnesses and determine the groundwater conditions.
2. Ground investigations should, ideally, be carried out in stages, involving desk studies, preliminary investigations and detailed investigations. Detailed investigations consist of test pitting, drilling and sampling, laboratory testing and *in situ* testing.
3. Often reasonable estimates can be made of the state and the undrained strength of soil in the ground from the geological history of deposition, erosion and groundwater changes. These estimates are, however, likely to be substantially modified by ageing (see Sec. 15.3).
4. The results of a ground investigation may be contained either in a factual report or in an interpretive report. The principal component of a factual report is the borehole logs which record all the details of each borehole: it will also record the procedures and results of the laboratory and *in situ* tests. An interpretive report should contain, in addition, cross-sections

Ground Investigations Ltd Borehole Log	Borehole No. A1

Contract:	Midfolk CC Highways Dept. Eastwich Bypass	Equipment and methods:
Location:	Chainage 2250 m	Light cable tool percussion,
Ground level:	+2 m OD	200 mm cased to 25 m: rotary
Date:	10/3/199x to 12/3/199x	core from 26 m

OD	Depth	Legend	Description	Groundwater observations	Samples Type	Samples Depth	Tests	
+2	0		Stiff grey silty CLAY with plant roots		D			
0	2		Soft grey CLAY with thin layers of silt and sand	Water level with casing at −8 m OD	TW 100	2.0 3.2	Vane	s_u 18 kPa
					TW 100	5.0 6.5	Vane	23 kPa
			As above becoming firm with depth		TW 100	8.0 9.3	Vane	35 kPa
−10	12		Dense brown fine to medium grained SAND with a little gravel	Water rose to 0 OD with casing at 12 to 22 m	D	13	SPT	N=32
					D	16	SPT	N=38
					D	18	SPT	N>50
			As above with increasing gravel		D	21		
−20	22		Stiff blue fissured CLAY (London Clay)	Hole dry	U 100	22.5 24		
−23	25		Weathered CHALK (weathering grades III to IV)	Standing water level −8 m OD		26.5		
					Core			
−32	34		End of hole			34		

Figure 16.8 Borehole log.

of the site showing all the principal soil and rock strata, recommended values for all the required design parameters, and, possibly, outline designs.

5. On completion of an investigation you should be able to provide, at least, the following information:

   (a) Cross-sections and plans showing the location of each of the principal strata and the groundwater conditions.

   (b) A list of the principal strata. This should include, for each stratum: descriptions of the nature and state of the soil or rock based on classification tests; the geological name (with capital letters); a description of the depositional environments and the subsequent geological events.

   (c) A full description of the groundwater conditions.

   (d) Values for the soil parameters required for the design: these would include the strength, stiffness and permeability (or consolidation) parameters appropriate to the ground conditions and the works.

   (e) Statements about the uncertainties (because you can never know everything about the ground from the results of a few boreholes and tests).

6. The variations of water content and undrained strength in the ground and their relationships to the Atterberg limits of plastic clays are very different for normally consolidated and overconsolidated clays.

7. The simple relationships linking the intrinsic parameters and soil profiles with soil classification tests and geological history are useful, particularly for preliminary design studies. However, we do not yet know enough about the fundamental mechanical properties of soils to select final design parameters from classification tests alone, so engineers must always conduct thorough ground investigations, including detailed laboratory and *in situ* testing.

REFERENCES

Clayton, C. R. I., N. E. Simons and M. C. Matthews (1982) *Site Investigation*, Granada, London.

Mair, R. J. and D. M. Wood (1987) 'Pressuremeter testing', in *CIRIA Ground Engineering Report*, Butterworth, London.

Meigh, A. C. (1987) 'Cone penetration testing', *CIRIA Ground Engineering Report*, Butterworth, London.

Skempton, A. W. (1957) 'Discussion on design and planning of the new Hong Kong airport', *Proc. ICE*, 7, 306.

FURTHER READING

BS 1377 (1991) *Methods of Test for Soils for Civil Engineering Purposes*, British Standards Institution, London.

BS 5930 (1981) *Code of Practice for Site Investigations*, British Standards Institution, London.

Clayton, C. R. I., N. E. Simons and M. C. Matthews (1982) *Site Investigation*, Granada, London.

Weltman, A. J. and J. M. Head (1983) *Site Investigation Manual*, CIRIA, London.

SEVENTEEN

STEADY STATE SEEPAGE

17.1 GROUNDWATER CONDITIONS

You know that water flows downhill and you have probably studied the flow of water in pipes and open channels in courses on hydraulics. Water also flows through soils in much the same way but now the flow is retarded as it flows past the grains. Theories for groundwater flow are covered in courses in hydraulics and all I will do here is consider the topics essential for geotechnical engineering. There are essentially three separate conditions for groundwater in geotechnical engineering and simple examples of these are illustrated in Fig. 17.1.

(a) Hydrostatic States

This condition, illustrated in Fig. 17.1(a), was discussed in Sec. 6.3. If the water table, or phreatic surface, is level there is no flow. Pore pressures are hydrostatic and are given by $u = \gamma_w h_w$.

(b) Steady State Seepage

If the phreatic surface is not level, as in Fig. 17.1(b), water will flow along flowlines such as ABC. At any point, such as at A, the pore pressures will be $u = \gamma_w h_w$, where h_w is the height of water in a standpipe. Note that the level of water in the pipe does not necessarily define the phreatic surface (see Sec. 17.5). Notice also that in Fig. 17.1(b) the flow is apparently uphill from A to B and that the pore pressure at C is greater than that at B.

The basic rule for the flow of water through a single element of soil is Darcy's law, which was introduced in Sec. 6.9 in connection with relative rates of loading and drainage. In this chapter we will extend Darcy's law to cover seepage through a whole region of soil. The essential feature of steady state seepage is that neither the pore pressures nor the rates of flow change with time. Since effective stresses remain constant the soil grains can be taken to be stationary as water flows through the pore channels.

(c) Consolidating Soil

When pore pressures change with time effective stresses and soil volumes also change with time. This process, which couples Darcy's seepage theory with soil compression and swelling is known

(a) Hydrostatic

(b) Steady state seepage

(c) Consolidating soil

Figure 17.1 Groundwater conditions.

as consolidation and was covered in Chapter 14. During consolidation, pore pressures are the sum of the steady state pore pressures $u_0 = \gamma_w h_{w0}$ and the excess pore pressure $\bar{u} = \gamma_w \bar{h}_w$, as shown in Fig. 17.1(c). Graphs of excess pore pressure $\bar{u}$ at given times are called isochrones.

17.2 PRACTICAL PROBLEMS OF GROUNDWATER FLOW

Any child who has dug a hole in the beach or constructed a small soil dam across a stream has soon recognized the importance of groundwater in ground engineering. It is impossible to excavate much below the groundwater table and dams soon fail by downstream erosion, even if they are not overtopped first. The hole can only be continued if water is pumped from the excavation, and possibly from the surrounding ground as well, and engineers will need to determine the quantities of water to be pumped. They will also be interested in the quantities of water leaking from water storage dams.

It is common knowledge that landslides occur most frequently after periods of rainfall when pore pressures in the ground are highest. (Remember that this has nothing to do with water lubricating soil.) We have already seen that soil strength and stiffness are governed by the effective stresses which depend on the pore pressures as well as on the total stresses, so that calculation of pore pressures in soil with steady state seepage will be an essential component of geotechnical design calculations.

Figure 17.2 illustrates two typical cases of steady state seepage in geotechnical problems. In both cases water flows from regions of high water level to regions of low water level along

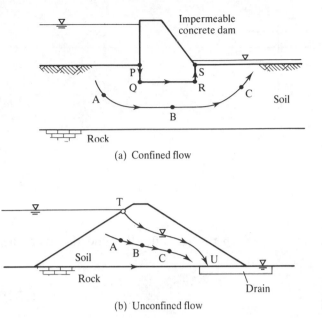

(a) Confined flow

T

Soil

Rock

Drain

(b) Unconfined flow

Figure 17.2 Problems in groundwater flow.

flowlines such as ABC: notice that in Fig. 17.2(a) the water apparently flows uphill from B to C. In Fig. 17.2(a) the flow is confined because the top flowline PQRS is confined by the impermeable concrete dam. In Fig. 17.2(b) the flow is unconfined and there is a phreatic surface, which is also the top flowline TU. In both cases we will be interested in calculating both the rates of leakage below or through the dams and the distributions of pore pressures.

In Fig. 17.2(a) water flows upwards in the region of C, where the flowline emerges at the downstream ground surface. If the seepage velocities are large, soil grains may be disturbed and washed away. If this should happen the erosion would seriously jeopardize the stability of the dam. The same thing might happen to the dam in Fig. 17.2(b) if the downstream drain is inadequate so that the top flowline TU emerges from the downstream face of the dam. After overtopping this is the most common cause of failure of seaside dams.

17.3 ESSENTIALS OF STEADY STATE SEEPAGE

Darcy's law governing flow of water through soil is very like Ohm's law for the flow of electricity through a conducting material, and an electrical flow model can be used to solve problems in groundwater seepage. In both cases a potential causes a current to flow against a resistance so that electrical conductivity is analogous to permeability. We have already seen that hydraulic potential is not the same as pore pressure and it is necessary to include a term to take account of the elevation.

To define potential it is necessary to have a datum as in Fig. 17.3(a). Since it is only changes of potential that matter the datum could be anywhere, but it is best to put it low down to avoid negative values of potential. From Fig. 17.3(a), the potential at A is

$$P = h_w + z = \frac{u}{\gamma_w} + z \tag{17.1}$$

(Note that this is simply Bernoulli's expression for total head since, in groundwater seepage, the velocity terms are small compared with the pressure and elevation terms.)

(a)   (b)   **Figure 17.3** Pore pressure and potential

In Fig. 17.3(b) the points A and B are δs apart on the same flowline and the hydraulic gradient between A and B is

$$i = -\frac{\delta P}{\delta s} \tag{17.2}$$

The negative sign is introduced into Eq. (17.2) so that the hydraulic gradient is positive in the direction of flow. (Note that in Fig. 6.11 and in Eq. (6.13) the hydraulic potential and the hydraulic gradient were defined in terms of h_w only. This was allowable in that case because the flowlines in Fig. 6.11 were horizontal and so the z term in Eq. (17.1) remains constant. From now on we will work with potentials and hydraulic gradients using Eqs (17.1) and (17.2), taking account of pore pressure and elevation terms.)

Figure 17.4 shows two flowlines AB and CD at an average distance δb apart. The points A and C have the same potential and so do the points B and D. The lines AD and BD are called equipotentials (because they are lines of equal potential) and the average distance between them is δs. Flowlines and equipotentials intersect at 90° as shown. (The proof of this is given in textbooks on hydraulics, which also show that flowlines and equipotentials are given by Laplace equations.)

Figure 17.4 represents two-dimensional seepage through isotropic soil in which the value of k is the same in all directions and through a slice of unit thickness normal to the page; all

Figure 17.4 Flowlines and equipotentials.

Figure 17.5 Seepage and flow velocities.

the discussion in this chapter is for two-dimensional seepage. The rate of flow (in cubic metres per second) between the two flowlines is δq and the mean seepage velocity (in metres per second) is

$$V = \frac{\delta q}{\delta b} \tag{17.3}$$

Darcy's law states that

$$V = ki \tag{17.4}$$

where k is the coefficient of permeability which has the units of velocity (in metres per second). Typical values of k for soils were given in Sec. 6.9. Remember that for coarse-grained soils $k > 10^{-2}$ m/s while for fine-grained soils $k < 10^{-8}$ m/s; these very large differences mean that coarse-grained soils with high permeability can act as drains while fine-grained soils with very low permeability can be used as nearly watertight barriers in dams.

Notice that the seepage velocity V given by Eq. (17.4) is not the velocity of a drop of water as it seeps through the pore spaces. From Fig. 17.5 the velocity of the drop of water is $V_w = \delta q / \delta w$, where δw is the area occupied by the pore spaces in an area of soil δb and

$$\frac{V}{V_w} = \frac{\delta w}{\delta b} = 1 - \frac{1}{v} \tag{17.5}$$

where v is the specific volume. This means that if you use dye or a tracer to examine groundwater flow you will measure V_w, which is not the same as the velocity given by Darcy's law in Eq. (17.4).

17.4 FLOW THROUGH A SIMPLE FLOWNET

Figure 17.4 shows the conditions of steady state seepage through a single element bounded by two flowlines and two equipotentials. The rate of flow through the element is given by Eqs (17.3) and (17.4) as

$$\delta q = \delta bki \tag{17.6}$$

If we can assign a value of potential to an equipotential we could calculate the pore pressures from Eq. (17.1), if necessary interpolating between the equipotentials. The flow through a whole region and the pore pressures throughout the region can be found by considering an assembly of elements called a flownet.

Figure 17.6 shows a simple flownet. The flowlines and equipotentials intersect orthogonally,

Figure 17.6 Flow through a simple flownet.

and if $\delta s = \delta b$ the flownet is square. There are four flowlines and so the number of flow channels, N_f, is three. The total rate of flow through the region is Δq and, making use of Eq. (17.6),

$$\Delta q = N_f \, \delta q = N_f \, \delta b k i \tag{17.7}$$

Because the flowlines are straight and parallel, the seepage velocity, and hence the hydraulic gradient, is constant and so the equipotentials are equally spaced as shown. There are six equipotentials and so the number of equipotential drops in the square flownet, N_d, is five; therefore, from Eq. (17.2),

$$i = -\frac{\delta P}{\delta s} = -\frac{\Delta P}{N_d \, \delta s} \tag{17.8}$$

Hence, from Eqs (17.7) and (17.8) the rate of flow through the whole flownet is

$$\Delta q = -k \frac{N_f}{N_d} \Delta P \tag{17.9}$$

where ΔP is the change of potential across the whole flownet.

Although Eq. (17.9) was derived for the simple flownet in Fig. 17.6 with straight flowlines and equipotentials, it is applicable to any flownet with curved elements provided that the elements are 'square' in the sense that the flowlines and equipotentials intersect orthogonally and the mean dimensions of each element are the same (i.e. $\delta s = \delta b$). Notice that the ratio N_f/N_d depends only on the geometry of the boundary of the flownet so that in Fig. 17.6 we could have $N_f = 6$ and $N_d = 10$ by halving the size of each element. If the values of potential, P_1 and P_2, at the inflow and outflow boundaries are known the values of potential can be found at any equipotential (because the drop in potential is the same across any element) and the pore pressure at any point within the flownet can be calculated from Eq. (17.1).

Figure 17.7 Flownet for steady state flow into a trench excavation.

17.5 FLOWNET FOR TWO-DIMENSIONAL SEEPAGE

A solution to any problem in two-dimensional steady state seepage can be found by drawing a square flownet. This must be a proper scale drawing with the correct boundary conditions and, for a particular geometry, there is only one set of flowlines and equipotentials that satisfies the boundary conditions. The solution gives the rate of flow from Eq. (17.9) and the distribution of pore pressure from Eq. (17.1). Techniques for constructing flownets by sketching, by electrical models and by numerical analysis are covered in textbooks on hydraulics. All I will do here is to find solutions to two simple cases to illustrate the general principles.

In Fig. 17.7 water seeps from a flooded ground surface into a trench supported by walls and which is pumped dry. The geometry is symmetric about the centre-line. The flow is confined so there is no phreatic surface. If a standpipe is placed with its tip just at the ground level, such as at G or at C, water will rise to the ground surface; therefore AG is an equipotential with value P_1 and similarly CF is an equipotential with value P_2. Any impermeable boundary, such as the wall and the rock surface, must be a flowline and so is the axis of symmetry; therefore, ABC and DEF are flowlines because flowlines cannot cross. A roughly sketched flownet is shown in Fig. 17.7(b). This satisfies the boundary conditions in Fig. 17.7(a); flowlines and equipotentials are orthogonal and each element is more or less 'square' with approximately equal length and breadth. For this flownet the total number of flow channels is $N_f = 8$ (i.e. four on each side of the centre-line) and the number of equipotential drops is $N_d = 10$.

In Fig. 17.8 water seeps through a soil embankment dam to a drain in the downstream toe. The flow is unconfined and there is a phreatic surface in a position approximately as shown by the broken line. If a standpipe is placed with its tip anywhere on the upstream face, water will rise to the reservoir level so AB is an equipotential with value P_1. Similarly, the drain CD is an equipotential with value P_2. The top of the impermeable rock AC is a flowline. The phreatic surface BE is not precisely located by the geometry of the dam alone but its position will be fixed by the geometry of the flownet. The phreatic surface is a flowline and, on the phreatic surface, the pore pressure is zero. A roughly sketched flownet is shown in Fig. 17.8(b). Again this satisfies the boundary conditions, flowlines and equipotential are orthogonal and each element is more or less 'square'. Notice that the equipotentials intersect the phreatic surface at

(a)

(b)

Figure 17.8 Flownet for steady state seepage through a dam.

equal vertical intervals (because $u = 0$ along the phreatic surface). For this flownet, $N_f = 2$ and $N_d = 5$.

The level of water in a standpipe does not necessarily rise to the phreatic surface. In Fig. 17.8(b) the tip of the standpipe is on the equipotential HJ. If the tip of a standpipe is on the phreatic surface at J the water remains at J and so the level of water in any standpipe on HJ must be at the level of J. For the standpipe at H the water rises not to K on the phreatic surface but to the level of J as shown.

These flownets can be used to calculate the rates of leakage into the trench excavation and through the dam using Eq. (17.9). Note that this contains the coefficient of permeability k and the accuracy of the solution will depend more on how well you can determine a value for k than on how well you can draw a flownet. The flownets can also be used to calculate pore pressures. You will need these to calculate the loads on the walls and props in Fig. 17.7 and the stability of the dam slopes in Fig. 17.8, but to calculate pore pressures the flownet must be accurately drawn. Notice that the geometry of a flownet and the pore pressures are independent of the value of coefficient of permeability k.

The flownets shown in Figs 17.7 and 17.8 were sketched by me very quickly using a soft pencil and a good rubber. They are a bit rough—not all the elements are properly 'square' and sometimes the flownets and equipotential do not intersect exactly orthogonally—but they are probably good enough for many design calculations. They could be improved by use of an electrical model or a numerical analysis. The important thing about my flownets is that they satisfy the boundary conditions and there are no fundamental inconsistencies. You should now go to a book on hydraulics and study flownets for other cases, particularly for flow into drains, wells and slots.

17.6 PIPING AND EROSION

As water flows through soil the potential drops and the drag on the soil grains results in an increase in effective stress in the direct of flow. If the flow is upwards these seepage stresses act

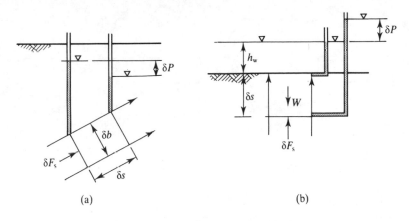

Figure 17.9 Seepage stresses and critical hydraulic gradient.

against the self-weight stresses and the resultant effective stresses reduce. This condition occurs in the base of the excavation in Fig. 17.7. If the upward flow is large the condition could occur where the effective stresses and the strength become zero, and this would clearly have very serious consequences for the stability of an excavation. This condition is known as piping, or boiling, and is the cause of quicksand: natural quicksands occur where there is an upward flow of water under artesian pressure.

Figure 17.9(a) shows flow through a single element of a flownet. The seepage force δF_s is due to the potential drop δP and, making use of Eq. (17.2),

$$\delta F_s = -\gamma_w \, \delta P \, \delta b = \gamma_w i \, \delta s \, \delta b \tag{17.10}$$

Dividing by δb, and for unit thickness normal to the page, the effective stress due to the seepage is

$$\delta \sigma'_s = \gamma_w i \, \delta s \tag{17.11}$$

which acts in the direction of the seepage flow.

Figure 17.9(b) shows the last element in a flownet where vertical upward seepage emerges at the ground surface. Note that since the flow is upward δs is measured negatively downwards. The stresses and pore pressures at a depth δs in the ground are

$$\sigma_v = -\gamma \, \delta s + \gamma_w h_w \tag{17.12}$$

$$u = \gamma_w(\delta P + h_w - \delta s) \tag{17.13}$$

Hence, making use of Eq. (17.2),

$$\sigma'_v = \gamma_w \, \delta s \left[\left(\frac{\gamma}{\gamma_w} - 1 \right) - i \right] \tag{17.14}$$

and the vertical effective stress σ'_v reduces with increasing i. If $\sigma'_v = 0$ the critical hydraulic gradient i_c is

$$i_c = \frac{\gamma}{\gamma_w} - 1 \tag{17.15}$$

For many soils γ is approximately 20 kN/m^3 and i_c is approximately unity. Piping or boiling will generally only occur for upward seepage towards the ground surface, as shown in Fig. 17.9(b). You can create piping in the apparatus shown in Fig. 6.6 by extending the standpipe and filling it to a height above ground level that is about twice the depth of the model.

Figure 17.10 Transformed flownet for anisotropic soil.

17.7 SEEPAGE THROUGH ANISOTROPIC SOILS

Many soils are layered either because they were naturally deposited in changing depositional environments or because they were compacted in layers, with the result that the permeability for horizontal flow k_h is often considerably greater than the permeability for vertical flow k_z. In this case the appropriate flownet is not square and flowlines and equipotentials do not intersect orthogonally, as shown in Fig. 17.10(a).

The flownet can, however, be made square by transforming the horizontal axis to H' and the mean coefficient of permeability to k', where

$$H' = \sqrt{\frac{k_z}{k_h}}\, H \tag{17.16}$$

$$k' = \sqrt{k_z k_h} \tag{17.17}$$

as shown in Fig. 17.10(b). The theoretical derivations for these transformations are given in textbooks on hydraulics and groundwater flow.

17.8 SUMMARY

1. For steady state seepage pore water pressures u at a point are given by the potential P:

$$P = h_w + z = \frac{u}{\gamma_w} + z \tag{17.1}$$

where z is the elevation of the point above an arbitrary datum.

2. Seepage of water through soil is governed by Darcy's law

$$V = ki \tag{17.4}$$

where V is the seepage velocity and i is the hydraulic gradient given by

$$i = -\frac{\delta P}{\delta s} \tag{17.2}$$

3. Steady state seepage through a region of soil is described by a square flownet consisting of an orthogonal net of flowlines and equipotentials. Pore pressures can be calculated from equipotentials. The total rate of flow through a flownet is given by

$$\Delta q = -k\frac{N_f}{N_d}\,\Delta P \tag{17.9}$$

where ΔP is the change of potential across the whole flownet. Flownets can be obtained by sketching orthogonal nets that satisfy the boundary conditions.

4. Seepage gives rise to seepage stresses which may cause instabilities due to piping or erosion. Seepage stresses (which are effective stresses in addition to the effective stresses due to unit weights) are given by

$$\delta \sigma'_s = \gamma_w i \, \delta s \qquad (17.11)$$

For upward seepage towards the ground surface the critical hydraulic gradient when $\sigma'_z = 0$ is given by

$$i_c = \frac{\gamma}{\gamma_w} - 1 \qquad (17.15)$$

5. For layered and anisotropic soils square flownets can be constructed that make use of a scale transformation and an equivalent permeability.

WORKED EXAMPLES

Example 17.1: Confined flow Figure 17.11 illustrates flow towards a long (out of the page) land drain through a layer of soil with permeability $k = 10^{-6}$ m/s sandwiched between clay and rock, both of which may be considered to be impermeable. The water level in the drain is 1 m below the water table 9 m away.

Figure 17.11

The phreatic surface is above the top of the soil and the flow is confined by the impermeable clay. A simple square flownet is shown in Fig. 17.11 in which $N_f = 3$ and $N_d = 9$. Taking the datum for potential at the rock level, $P_1 = 5$ m and $P_2 = 4$ m and, from Eq. (17.9) the rate of flow into the drain (from one side) per unit length out of the page is

$$\Delta q = -k \frac{N_f}{N_d} \Delta P = -10^{-6} \times \tfrac{9}{3} \times (4 - 5) = 3 \times 10^{-6} \text{ m}^3/\text{s}$$

At the point A the elevation is $z_a = 2$ m and the potential is

$$P_a = P_1 - \tfrac{3}{9} \Delta P = 5 - \tfrac{3}{9}(5 - 4) = 4.67 \text{ m}$$

Hence, from Eq. (17.1), the pore pressure at A is

$$u_a = \gamma_w (P_a - z_a) = 9.81 \times (4.67 - 2) = 26 \text{ kPa}$$

Example 17.2: Unconfined flow   Figure 17.12 illustrates leakage from a canal into a nearby river. (Both the river bank and the canal bank are supported by sheet piles that leak.) The coefficient of permeability of the soil is 10^{-6} m/s.

Figure 17.12

The phreatic surface joins the water levels in the river and canal and the flow is unconfined. From the flownet sketched $N_f = 3$ and $N_d = 7$ and, taking the datum for potential at the bed of the canal, $P_1 = 4$ m and $P_2 = 2$ m. Hence, from Eq. (17.9) the rate of leakage per unit length of the page is

$$\Delta q = -k \frac{N_f}{N_d} \Delta P = -10^6 \times \tfrac{7}{3} \times (2 - 4) \approx 5 \times 10^{-6} \text{ m/s}$$

At the point A, scaling from the diagram, the elevation is $z_a = 1.73$ m and the potential is

$$P_a = P_1 - \tfrac{5}{7} \Delta P = 4 - \tfrac{5}{7}(4 - 2) = 2.57 \text{ m}$$

Hence, from Eq. (17.1), the pore pressure at A is

$$u_a = \gamma_w(P_a - z_a) = 9.81 \times (2.57 - 1.73) = 8.2 \text{ kPa}$$

Notice that water in a standpipe at A rises to the level where the equipotential through A meets the phreatic surface.

FURTHER READING

Atkinson, J. H. and P. L. Bransby (1978) *The Mechanics of Soils*, McGraw-Hill, London.
Cedergren, H. R. (1967) *Seepage, Drainage and Flownets*, Wiley, New York.
Harr, E. (1962) *Groundwater and Seepage*, McGraw-Hill, New York.
Powers, J. P. (1992) *Construction Dewatering*, Wiley, New York.
Somerville, S. H. (1986) *Control of Ground Water for Temporary Works*, CIRIA, Report 113, London.

CHAPTER EIGHTEEN

STABILITY OF SOIL STRUCTURES USING BOUND METHODS

18.1 INTRODUCTION

In Chapters 8 to 12 I considered the behaviour of single elements of soil, either in the ground or in laboratory tests, and I developed simple theories for strength of soil and simple constitutive equations relating increments of stress and strain. What we have to do now is to apply these theories to the behaviour of geotechnical structures such as foundations, slopes and retaining walls. As discussed earlier, solutions for problems in mechanics must satisfy the three conditions of equilibrium, compatibility and material properties. It is fairly obvious that complete solutions, satisfying these conditions with the material properties for soil, will be very difficult to obtain, even for very simple foundations and slopes.

First, I will consider the conditions of ultimate collapse where the important material property is the soil strength. Remember that, as always, it is necessary to distinguish between cases of undrained and drained loading. For undrained loading the ultimate strength of soil is given by

$$\tau = s_u \tag{18.1}$$

where s_u is the undrained strength. For drained loading where pore pressures can be determined from hydrostatic groundwater conditions or from a steady state seepage flownet the strength is given by

$$\tau' = \sigma' \tan \phi_c' \tag{18.2}$$

where ϕ_c' is the critical state friction angle. These strengths give the ultimate collapse states. To design safe structures or to limit ground movements, they can be reduced by a factor F_s, as described in later sections.

Even with these relatively simple expressions for soil strength it is still quite difficult to obtain complete solutions and the standard methods used in geotechnical engineering involve simplifications. There are two basic methods: the bound methods described in this chapter and the limit equilibrium method described in the next chapter. Both methods require approximations and simplifications which will be discussed in due course.

18.2 THEOREMS OF PLASTIC COLLAPSE

In order to simplify stability calculations it is possible to ignore some of the conditions of equilibrium and compatibility and to make use of important theorems of plastic collapse. It turns out that by ignoring the equilbrium condition you can calculate an upper bound to the collapse load so that if the structure is loaded to this value it must collapse; similarly, by ignoring the compatibility condition you can calculate a lower bound to the collapse load so that if the structure is loaded to this value it cannot collapse. Obviously the true collapse load must lie between these bounds.

The essential feature of the upper and lower bound calculations is that rigorous proofs exist which show that they will bracket the true collapse load. Thus, although the two methods of calculation have been simplified by ignoring, for the first, equilibrium and, for the second, compatibility, no major assumptions are needed (other than those required to prove the bound theorems in the first place). What has been lost by making the calculations simple is certainty; all you have are upper and lower bounds and you do not know the true collapse load (unless you can obtain equal upper and lower bounds). Usually you can obtain upper and lower bounds that are fairly close to one another so the degree of uncertainty is quite small.

I am not going to prove the plastic collapse theorems here and I will simply quote the results. A condition required to prove the theorems is that the material must be perfectly plastic. This means that, at failure, the soil must be straining at a constant state with an associated flow rule so that the vector of plastic strain increment is normal to the failure envelope (see Chapter 3). The first condition, straining at a constant state, is met by soils at their ultimate or critical states, given by Eqs (18.1) and (18.2). The second condition is illustrated in Fig. 18.1(a) for undrained loading and in Fig. 18.1(b) for drained loading.

In both cases elastic strains must be zero since the stresses remain constant; thus total and plastic strains are the same. For undrained loading the failure envelope given by Eq. (18.1) is horizontal and the volumetric strains are zero (because undrained means constant volume) and so the vector of plastic strain $\delta\varepsilon^p$ is normal to the failure envelope as shown. For drained loading the failure envelope is given by Eq. (18.2) and if the flow rule is associated the angle of dilation

(a) Undrained loading   (b) Drained loading

Figure 18.1 Straining of perfectly plastic soil with an associated flow rule.

at the critical state ψ_c is

$$-\frac{\delta\varepsilon_n^p}{\delta\gamma^p} = \tan\psi_c = \tan\phi_c' \tag{18.3}$$

At the critical state, however, soil strains at a constant state (i.e. at a constant volume) and so $\psi_c = 0$, which means that, at failure at the critical state, the flow rule is not associated and soil in drained loading is not perfectly plastic. This does not actually matter very much as you can prove that an upper bound for a material with $\psi_c = \phi_c'$ is still an upper bound, even if ψ_c is less than ϕ_c', but you can not do the same for the lower bound. In practice upper and lower bounds for soil structures calculated with $\psi_c = \phi_c'$ give good agreement with experimental observations and, although the lower bound solution is not absolutely rigorous, the errors seem to be small.

The statements of the bound theorems are simple and straightforward:

1. Upper bound. If you take any compatible mechanism of slip surfaces and consider an increment of movement and if you show that the work done by the stresses in the soil equals the work done by the external loads, the structure must collapse (i.e. the external loads are an upper bound to the true collapse loads).
2. Lower bound. If you can determine a set of stresses in the ground that are in equilibrium with the external loads and do not exceed the strength of the soil, the structure cannot collapse (i.e. the external loads are a lower bound to the true collapse loads).

To calculate an upper bound you must satisfy the conditions of compatibility and the material properties (which govern the work done by the stresses in the soil), but nothing is said about equilibrium. To calculate a lower bound you must satisfy the conditions of equilibrium and the material properties (which determine the strength), but nothing is said about displacements or compatibility. Because a structure with an upper bound load must collapse this is often known as the unsafe load and because a structure with a lower bound load cannot collapse this is known as the safe load. The basic principles of these upper and lower bound methods are also used to calculate stability of framed structures by using plastic hinges to create mechanisms or by using elastic analysis to calculate yield stresses at critical sections.

In the present context the terms upper and lower bounds have the very specific meanings associated with the bound theorems. Engineers also investigate bounds to structural behaviour by investigating the consequences of optimistic and pessimistic values for material properties, but bounds calculated in this way are obviously quite different from the present meaning.

18.3 COMPATIBLE MECHANISMS OF SLIP SURFACES

To calculate an upper bound a mechanism of slip surfaces must meet the requirements of compatibility. These requirements determine both the allowable shape of individual slip surfaces and their general arrangement.

Figure 18.2(b) shows a segment of a curved slip surface represented by a double line and Fig. 18.2(a) shows an enlarged small element. On one side the material is stationary and on the other side there is an increment of displacement δw at an angle ψ. The length along the slip surface is constant so it is a zero extension line (see Sec. 2.6). From Eq. (2.11) and from the geometry of Fig. 2.9, slip surfaces makes angles α and β to the major principle planes where

$$\alpha = \beta = 45° + \tfrac{1}{2}\psi \tag{18.4}$$

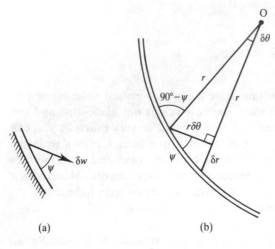

(a)   (b)   **Figure 18.2** Geometry of a slip plane.

From the geometry of Fig. 18.2(b),

$$\frac{dr}{r\,d\theta} = \tan\psi \tag{18.5}$$

and therefore

$$\frac{r_B}{r_A} = \exp(\Delta\theta \tan\psi) \tag{18.6}$$

where $\Delta\theta$ is the angle between the radii r_A and r_B. This is the equation of a logarithmic spiral for $\psi > 0$ but, for undrained loading when $\psi = 0$,

$$\frac{r_B}{r_A} = \exp(0) = 1 \tag{18.7}$$

This is the equation of a circular arc. Also, as $r_A \to \infty$, Eqs (18.6) and (18.7) tend to the equation for a straight line. Thus, for drained loading where $\psi = \phi_c'$, slip surfaces may be straight lines or logarithmic spirals while, for undrained loading where $\psi = 0$, slip surfaces may be straight lines or circular arcs. In Fig. 18.2(b) the radii intersect the curved slip surface at a constant angle $(90° - \psi)$ and hence radii may also be slip surfaces.

Slip surfaces can be assembled to form a compatible mechanism of plastic collapse; a number of simple mechanisms are illustrated in Fig. 18.3. These may consist of straight lines or curves (circular arcs for undrained loading with $\psi_c = 0$ or logarithmic spirals for drained loading with $\psi_c = \phi_c'$) or combinations of straight lines and curves. Notice that in Fig. 18.3(f) the curved section is in fact a fan with a radial slip surface and these are required to make the mechanism compatible by constructing a displacement diagram as described in Sec. 2.3.

18.4 WORK DONE BY INTERNAL STRESSES AND EXTERNAL LOADS

To determine an upper bound it is necessary to calculate the work done by the internal stresses and by the external loads during an increment of movement of a compatible mechanism. The work done by a force is simply the product of the force and the increment of displacement resolved into the direction of the force. We can always determine the increments of displacements, resolved in any direction, from a displacement diagram.

External loads arise from concentrated forces from small foundations, from distributed

(a)   (b)   (c)

(d)   (e)

(f)

Figure 18.3 Compatible mechanisms.

Figure 18.4 Work done by external loads.

(a)   (b)

Figure 18.5 Work done by internal stresses on slip planes.

stresses below embankments and wide foundations and from the self-weight of the soil. External loads from concentrated forces are easy to determine and are the same for drained and for undrained loading, but for distributed stresses and self-weight drained and undrained loading must be considered separately. Figure 18.4 shows an element of soil with unit weight γ and with a total stress p and a concentrated load F at the top surface where the pore pressure is u. There is an increment of displacement δw in the direction of the surface stress, the concentrated load and self-weight forces. For undrained loading the increment of work δE is

$$\delta E = F \, \delta w + pA \, \delta w + \gamma V \, \delta w \tag{18.8}$$

For drained loading the water remains stationary so the work is done by the effective stresses only; hence

$$\delta E = F \, \delta w + (p - u)A \, \delta w + (\gamma - \gamma_w)V \, \delta w \tag{18.9}$$

For dry soil simply put $u = \gamma_w = 0$ in Eq. (18.9).

The work done by the internal stresses is the work dissipated by plastic straining in the material in the thin slip surfaces that make up the compatible mechanism and, again, undrained and drained loading must be considered separately. Figure 18.5 shows short lengths of slip surfaces that have increments of displacement δw as shown. Since the soil is at the critical state in each case the stresses are given by Eqs (18.1) and (18.2) and, for drained loading, the shear and normal strains are related by Eq. (18.3).

In Fig. 18.5(b) for drained loading the water remains stationary, the work is done by the effective stresses and hence

$$\delta W = \tau' L \, \delta l - \sigma_n' L \, \delta n \tag{18.10}$$

Note that for dilation the work done by the normal stress is negative since σ_n' and δn are in opposite directions. From Eq. (18.10), with the volume of the slip plane $V = Ly$,

$$\delta W = V(\tau' \, \delta\gamma + \sigma_n' \, \delta\varepsilon_n) = V\tau' \, \delta\gamma\left(1 - \frac{\tan \psi_c}{\tan \phi_c'}\right) \tag{18.11}$$

However, for a perfectly plastic material $\psi_c = \phi_c'$ and so the work dissipated by the internal stresses for drained loading is

$$\delta W = 0 \tag{18.12}$$

This is a very surprising result and presents difficulties which I will not explore here. The implication is that a perfectly plastic frictional material is both dissipative and conservative, which is nonsense. The conclusion must be that the flow rule for a frictional material cannot be associated. Nevertheless, the result given by Eq. (18.12) is very convenient and it may be used to calculate upper bounds for frictional materials like soil.

In Fig. 18.5(a) for undrained loading the increment of work done by the total stresses τ and σ is

$$\delta W = \tau L \, \delta w = s_u L \, \delta w \tag{18.13}$$

Note that for undrained or constant volume straining no work is done by the normal stress σ_n because there is no displacement normal to the slip surface. For an upper bound calculation you must evaluate Eq. (18.13) for all the slip planes in the compatible mechanism.

18.5 SIMPLE UPPER BOUNDS FOR A FOUNDATION

In order to illustrate the use of the bound theorems I shall obtain solutions for the bearing capacity of a foundation subject to undrained loading. Figure 18.6 shows a foundation with unit length out of the page so that the width B is equal to the area A. As the foundation load F and bearing pressure q are raised the settlement ρ will increase until the foundation can be said to have failed at the collapse load F_c or the bearing capacity q_c. The foundation is smooth so there are no shear stresses between the soil and the foundation. I will obtain solutions using, firstly,

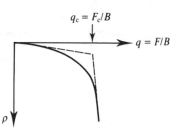

Figure 18.6 Bearing capacity of a simple foundation.

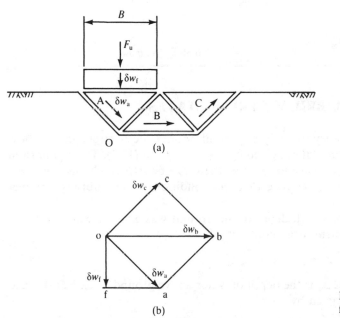

Figure 18.7 Mechanism of collapse for a foundation.

a simple mechanism and, secondly, two stress discontinuities, and later I will obtain more complex solutions using a slip fan and a stress fan. The purpose here is to illustrate the principles of the bound solutions; I will consider the bearing capacity of foundations in more detail in Chapter 22.

Figure 18.7(a) shows a simple mechanism consisting of three triangular wedges and Fig. 18.7(b) is the corresponding displacement diagram. The increments of work done by the self-weight forces sum to zero since block B moves horizontally while the vertical components of the displacements of blocks A and C are equal and opposite. Hence, from Eq. (18.8), we have

$$\delta E = F_u \, \delta w_f \tag{18.14}$$

In order to calculate the work done by the internal stresses on the slip planes, from Eq. (18.13) it is easiest to tabulate s_u, L and δw for each slip plane. Hence, from Table 18.1,

$$\delta W = 6 s_u B \, \delta w_f \tag{18.15}$$

and, equating δE and δW, an upper bound for the collapse load is

$$F_u = 6 B s_u \tag{18.16}$$

Table 18.1

Slip plane	Shear stress	Length	Displacement	$\delta W = s_u L\, \delta w$
oa	s_u	$\dfrac{1}{\sqrt{2}}B$	$\sqrt{2}\,\delta w_f$	$s_u B\, \delta w_f$
ob	s_u	B	$2\delta w_f$	$2s_u B\, \delta w_f$
oc	s_u	$\dfrac{1}{\sqrt{2}}B$	$\sqrt{2}\,\delta w_f$	$s_u B\, \delta w_f$
ab	s_u	$\dfrac{1}{\sqrt{2}}B$	$\sqrt{2}\,\delta w_f$	$s_u B\, \delta w_f$
bc	s_u	$\dfrac{1}{\sqrt{2}}B$	$\sqrt{2}\,\delta w_f$	$s_u B\, \delta w_f$
fa	0	B	δw_f	0
			Total	$\overline{6s_u B\, \delta w_f}$

18.6 DISCONTINUOUS EQUILIBRIUM STRESS STATES

To calculate a lower bound it is necessary to analyse an equilibrium state of stress and to show that it does not exceed one of the failure criteria given by Eqs (18.1) and (18.2). The equilibrium states of stress may vary smoothly from place to place or there can be sudden changes of stress across stress discontinuities, provided, of course, that the conditions of equilibrium are met across the discontinuities.

The variation of vertical total stress with depth in the ground was given in Sec. 6.2. From Fig. 18.8 the vertical stress on an element at a depth z is

$$\sigma_v = \gamma z + q + \gamma_w z_w \tag{18.17}$$

where q is a uniform surface stress and z_w is the depth of water above ground level. For drained loading the effective vertical stress is given by

$$\sigma_v' = \sigma_v - u \tag{18.18}$$

where u is the (steady state) pressure.

In Fig. 18.9(a) there are two regions A and B separated by a discontinuity represented by a single bold line; the stresses in each region are uniform and are characterized by the magnitudes and directions of the major principal stresses σ_{1a} and σ_{1b} as shown. The rotation in the direction of the major principle stress across the discontinuity is $\delta\theta = \theta_b - \theta_a$. The Mohr circles of total stress are shown in Fig. 18.9(b). The point C represents the normal and shear stresses on the discontinuity and the poles of the circles are found by drawing $P_a - C - P_b$ parallel to the

Figure 18.8 Vertical stress in the ground.

Figure 18.9 Change of stress across a discontinuity.

discontinuity in Fig. 18.9(a). Hence the directions of the major principal planes are given by the broken lines in Fig. 18.9(b) and, from the properties of the Mohr circle construction given in Sec. 2.4, we can mark $2\theta_a$ and $2\theta_b$, the angles subtended by σ_{1a} and σ_{1b}, and the normal stress on the discontinuity.

As usual it is necessary to consider undrained and drained loading separately. Figure 18.10 shows the analysis for undrained loading. Both Mohr circles of total stress touch the failure line given by Eq. (18.1). From the geometry of Fig. 18.10(b), noting that $AC = s_u$,

$$\delta s = 2s_u \sin \delta\theta \qquad (18.19)$$

Hence the change of total stress across a discontinuity is simply related to the rotation $\delta\theta$ of the direction of the major principal stress.

Figure 18.11 shows the analysis for drained loading. Both Mohr circles of effective stress touch the failure line given by Eq. (18.2) and the angle ρ' defines the ratio τ_n'/σ_n' on the discontinuity. It is convenient to define an angle P as shown in Fig. 18.12, where

$$P = 90° - \delta\theta \qquad (18.20)$$

From the geometry of Fig. 18.12, noting that $A'C' = t_a'$,

$$\sin P = \frac{A'D'}{t_a'} \qquad \sin \rho' = \frac{A'D'}{s_a'} \qquad (18.21)$$

(a)

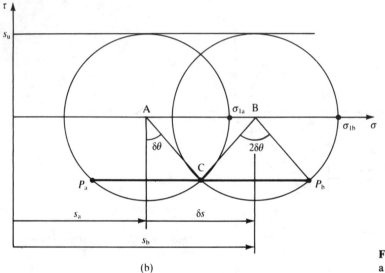

(b)

Figure 18.10 Change of stress across a discontinuity for undrained loading.

Hence, making use of Eq. (18.20),

$$\sin \rho' = \sin P \sin \phi'_c = \cos \delta\theta \sin \phi'_c \tag{18.22}$$

With the aid of the constructions in Fig. 18.12 and noting that $O'E' = O'F'$,

$$\sin(P + \rho') = \frac{O'E'}{s'_a} \qquad \sin(P - \rho') = \frac{O'F'}{s'_b} \tag{18.23}$$

and hence, making use of Eq. (18.20),

$$\frac{s'_b}{s'_a} = \frac{\cos(\delta\theta - \rho')}{\cos(\delta\theta + \rho')} \tag{18.24}$$

where ρ' is given by Eq. (18.22).

From Eqs (18.24) and (18.22) the change of effective stress across a discontinuity is simply related to the rotation $\delta\theta$ of the direction of the major principal stress.

(a)

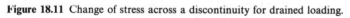

(b)

Figure 18.11 Change of stress across a discontinuity for drained loading.

Figure 18.12

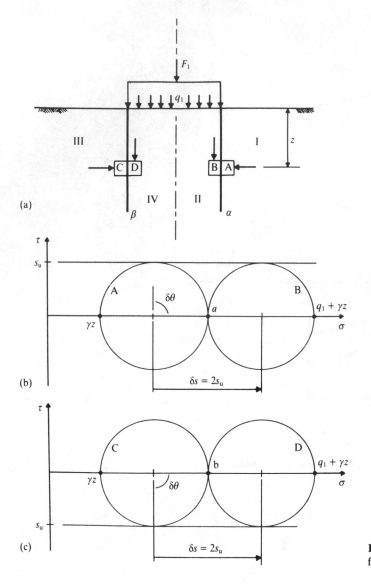

Figure 18.13 Equilibrium stress field for a foundation.

18.7 SIMPLE LOWER BOUNDS FOR A FOUNDATION

We can now obtain a simple lower bound solution for the foundation shown in Fig. 18.5. Figure 18.13(a) shows a state of stress with two vertical stress discontinuities where the state of stress is symmetric about the centre-line. Shear stresses on horizontal and vertical planes are zero and hence, from Eq. (18.17), the vertical stresses in elements A and C in regions I and III are

$$\sigma_z = \gamma z \tag{18.25}$$

and the vertical stresses in elements B and D in regions II and IV are

$$\sigma_z = q_1 + \gamma z \tag{18.26}$$

Figure 18.13(b) shows the Mohr circles of total stress for the elements A and B and Fig. 18.13(c) shows the circles for the elements C and D; the points a and b represent the stresses on the discontinuities marked α and β in Fig. 18.7(a). From the geometry of Figs 18.13(b) and (c),

$$q_1 + \gamma z = 4s_u + \gamma z \tag{18.27}$$

and hence a lower bound for the collapse load is

$$F_1 = 4s_u B \tag{18.28}$$

Alternatively, we could consider the rotations of the directions of the major principle stresses across the discontinuities, making use of Eq. (18.19). For each discontinuity $\delta\theta = 90°$ and $\delta s = 2s_u$; hence, from the geometry of Fig. 18.13(b) and (c) we obtain Eqs (18.27) and (18.28).

The mean of the upper and lower bound solutions gives $F_c = 5s_u$ and the bounds differ by about ± 20 per cent from this mean. Bearing in mind the problems in determining true values of s_u for natural soils, which may not be either isotropic or homogeneous, these simple bounds may be adequate for simple routine designs. However, in order to illustrate the use of slip fans and stress fans we will examine some alternative solutions.

18.8 UPPER AND LOWER BOUND SOLUTIONS USING FANS

In Fig. 18.3(f) there is a combination of straight and curved slip surfaces and in order to have a compatible mechanism it is necessary to have a fan a slip surfaces as illustrated. Figure 18.14 shows mechanisms and displacement diagrams for slip fans: Fig. 18.14(a) is for undrained loading and Fig. 18.14(b) is for drained loading. You should work your way through these together with the description of the construction of displacement diagrams given in Sec. 2.3. From the geometry of Fig. 18.14(a),

$$r_b = r_a \quad \text{and} \quad \delta w_b = \delta w_a \tag{18.29}$$

and so the radius of the fan and the increment of displacement remain constant through a slip fan for undrained loading. From the geometry of Fig. 18.14(b),

$$r_b = r_a \exp(\theta_f \tan \psi) \tag{18.30}$$

$$\delta w_b = \delta w_a \exp(\theta_f \tan \psi) \tag{18.31}$$

where θ_f is the fan angle; thus the outer arcs of the slip fan and the displacement diagram are both logarithmic spirals.

For a slip fan like that shown in Fig. 18.14(a), it is necessary to evaluate the work done on the circular slip surface and on all the radial slip surfaces. From Fig. 18.15, summing for the elements of the circular arc and for the radial slip surfaces, the increment of work done by the internal stresses through the fan is

$$\delta W = \sum s_u R(\delta w \, \delta\theta) + \sum s_u (R \, \delta\theta) \, \delta w \tag{18.32}$$

Hence, in the limit,

$$\delta W = 2s_u R \, \delta w \int_0^{\theta_f} d\theta \tag{18.33}$$

and

$$\delta W = 2s_u R \, \delta w \, \theta_f = 2s_u R \, \delta w \, \Delta\theta \tag{18.34}$$

where θ_f is the fan angle which is equal to the change $\Delta\theta$ in the direction of the vector of displacement δw across slip fan.

We can also consider the change of stress from one region to another across a fan of discontinuities, as shown in Fig. 18.16. (The fan of stress discontinuities in Fig. 18.16 is not

(a) Undrained loading

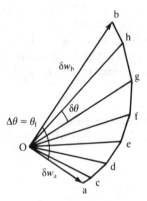

(b) Drained loading

Figure 18.14 Slips fans and corresponding displacement diagrams.

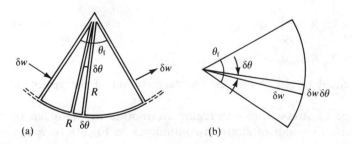

(a)

(b)

Figure 18.15 Work done in a slip fan.

Figure 18.16 Rotation of the direction of the major principal stress across a stress fan.

necessarily the same as the fan of slip surfaces in Fig. 18.14.) The fan angle θ_f is equal to the rotation $\Delta\theta$ of the direction of the major principal stress across the fan. Figure 18.17(a) shows a stress fan for undrained loading and Fig. 18.17(b) shows the Mohr circles of total stress for the outermost discontinuities; within the fan there are a great many radial discontinuities and there are equally a great many Mohr circles between those shown. Note that the outermost limits of the fan are defined by $\theta_a = \theta_b = 45°$. From Eq. (18.19), as $\delta\theta \to 0$,

$$\frac{ds}{d\theta} = 2s_u \tag{18.35}$$

Figure 18.17 Change of stress across a stress fan for undrained loading.

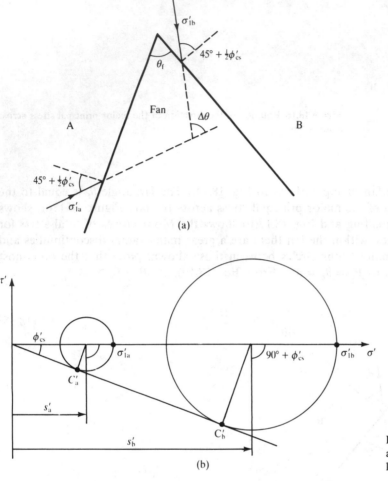

Figure 18.18 Change of stress across a stress fan for drained loading.

and integrating through the fan from region A to region B,

$$\Delta s = 2s_u \, \Delta\theta = 2s_u \, \theta_f \tag{18.36}$$

Figure 18.18 shows a stress fan and the corresponding Mohr circles for drained loading. As before there will be a great number of additional radial discontinuities and Mohr circles between the outermost ones. Note that the limits of the fan zone are defined by $\theta_a = \theta_b = 45° + \frac{1}{2}\phi'_c$. From Eq. (18.24), the change of stress across a discontinuity can be written as

$$\frac{ds'}{s'} = \frac{2 \sin \delta\theta \sin \rho'}{\cos(\delta\theta + \rho')} \tag{18.37}$$

As $\delta\theta \to 0$, from Eq. (18.22) we have $\rho' = \phi'_c$ and from Eq. (18.37),

$$\frac{ds'}{d\theta} = 2s' \tan \phi'_c \tag{18.38}$$

Hence, integrating through the fan from region A to region B,

$$\frac{s'_b}{s'_a} = \exp(2 \tan \phi'_c \, \Delta\theta) = \exp(2 \tan \phi'_c \, \theta_f) \tag{18.39}$$

Equations (18.36) and (18.39) give the changes of stress across stress fans in terms of the soil strength s_u or ϕ'_c and the fan angle θ_f or the rotation $\Delta\theta$ of the direction of the major principal stress.

18.9 BOUND SOLUTIONS FOR THE BEARING CAPACITY OF A FOUNDATION USING FANS

The simple upper and lower bound solutions obtained earlier can now be modified by adding slip fans or stress fans.

(a) Upper Bound with a Slip Fan

Figure 18.19(a) shows a mechanism consisting of two triangular wedges and a slip fan and Fig. 18.19(b) is the corresponding displacement diagram. As before the work done by the self-weight forces sums to zero and, from Eq. (18.8),

$$\delta E = F_u \, \delta w_f \tag{18.40}$$

The radius of the fan is $R = B/\sqrt{2}$, the fan angle is $\theta_f = \frac{1}{2}\pi$ and $\delta w_a = \sqrt{2} \, \delta w_f$. Hence, from Eq. (18.34) the work done by the internal stresses in the slip fan is

$$\delta W = 2s_u R \, \delta w \, \theta_f = \pi s_u B \, \delta w_f \tag{18.41}$$

and, from Table 18.2, for the whole mechanism

$$\delta W = (2 + \pi)s_u B \, \delta w_f \tag{18.42}$$

Equating δE and δW, the upper bound for the collapse load is

$$F_u = (2 + \pi)Bs_u \tag{18.43}$$

(b) Lower Bound with Stress Fans

Figure 18.20(a) shows a state of stress with two stress fans in regions II and IV. As before, the state of stress is symmetric about the centre-line and Eqs (18.25) and (18.26) apply in regions I

Figure 18.19 Mechanism of collapse for a foundation.

Table 18.2

Slip plane	Shear stress	Length	Displacement	$\delta W = s_u L \, \delta w$
oa	s_u	$\dfrac{1}{\sqrt{2}} B$	$\sqrt{2}\,\delta w_f$	$s_u B \, \delta w_f$
ob	s_u	$\dfrac{1}{\sqrt{2}} B$	$\sqrt{2}\delta w_f$	$2s_u B \, \delta w_f$
Fan	s_u	—	—	$\pi s_u B \, \delta w_f$
fa		B	δw_f	0
			Total	$(2 + \pi)s_u B \, \delta w_f$

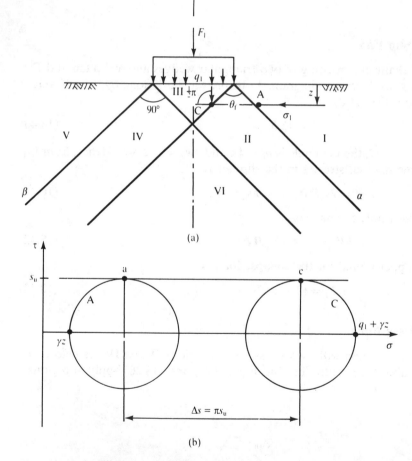

Figure 18.20 Equilibrium stress field for a foundation.

and III respectively. Figure 18.20(b) shows Mohr circles of total stress for elements at A and C and the points a and c represent the stresses on the outermost discontinuities in the fan in region II. From the geometry of Fig. 18.20, the fan angle is $\theta_f = 90° = \pi/2$ and from Eq. (18.36) the change of stress through the fan is

$$\Delta s = s_u \, \Delta\theta_f = \pi s_u \qquad (18.44)$$

From the geometry of Fig. 18.20(b),

$$q_1 + \gamma z = (2 + \pi)s_u + \gamma z \tag{18.45}$$

and hence a lower bound for the collapse load is

$$F_1 = (2 + \pi)Bs_u \tag{18.46}$$

Strictly, we should examine the state of stress in region VI where the stress fans overlap, but the analysis is lengthy and beyond the scope of the present book. It is intuitively fairly clear that the stresses in region VI will be less critical than those near the edges of the foundation and that the conditions in the overlapping stress fans will tend to cancel each other out.

Notice that the upper and lower bounds given by Eqs (18.43) and (18.46) are equal and so they must be an exact solution. We have been very fortunate to obtain an exact solution with such simple upper and lower bound solutions; normally you would only be able to obtain unequal bounds.

18.10 SUMMARY

1. Estimates of the collapse of structures can be found from relatively simple upper and lower bound calculations. An upper bound solution gives an unsafe load and if this load is applied the structure must collapse; a lower bound gives a safe load and with this load the structure cannot collapse.
2. To calculate an upper bound you have to choose a compatible mechanism of collapse and equate the work done by the external loads with the work done by the internal stresses. Mechanisms consists of slip surfaces that have circular arcs, logarithmic spirals or straight lines and may be arranged as fan zones.
3. To calculate a lower bound you need to find a distribution of stress that is in equilibrium with the external loads and does not exceed the appropriate failure criterion. An equilibrium state of stress may have strong discontinuities or stress fans.

The cases discussed in this chapter have been relatively simple and were intended simply to illustrate the basic principles of the upper and lower bound calculations. They concentrated on undrained loading and only considered cases of smooth foundations with vertical loads so that the shear stress between the soil and the foundation was zero. Other, more complicated, cases for drained loading and for rough walls and foundations are given by Atkinson (1981).

WORKED EXAMPLES

Example 18.1: Loads on trench struts for undrained soil The trench shown in Fig. 18.21 is supported by smooth sheet piles held apart by struts, 1 m apart out of the page, placed so that the piles do not rotate.

Figure 18.21

(a)　　　　　　　　　　　　　　　　(b)　　　　**Figure 18.22**

(a) Upper bound. Figure 18.22(a) shows a collapse mechanism and Fig. 18.22(b) is the corresponding displacement diagram. The forces acting on the moving block (for a slice 1 m thick out of the page) are

$$Q = qH = 80 \times 5 = 400 \,\text{kN}$$

$$W = \tfrac{1}{2}\gamma H^2 = \tfrac{1}{2} \times 20 \times 5^2 = 250 \,\text{kN}$$

From the displacement diagram, for $\delta v = 1$,

$$\delta v = \delta h = 1 \quad \text{and} \quad \delta w = \sqrt{2}$$

Hence, from Eq. (18.8), the work done by the external forces is

$$\delta E = Q\,\delta v + W\,\delta v - P_{\text{u}}\,\delta h = 400 + 250 - P_{\text{u}}$$

From Eq. (18.13), the work dissipated in the slip plane with length $5\sqrt{2}$ m is

$$\delta W = s_{\text{u}} L\,\delta w = 40 \times 5\sqrt{2} \times \sqrt{2} = 400$$

Hence, equation $\delta E = \delta W$,

$$P_{\text{u}} = 400 + 250 - 400 = 250 \,kN$$

(b) Lower bound. Figure 18.23(a) shows a typical element in an equilibrium stress field and Fig. 18.23(b) is the corresponding Mohr circle of total stress. From these,

$$\sigma_z = q + \gamma z = 80 + 20z$$

$$\sigma_h = \sigma_z - 2s_{\text{u}} = (80 + 20z) - (2 \times 40) = 20z$$

(a)　　　　　　　　　　　　　　　　(b)

Figure 18.23

Hence, integrating over the height of the trench,

$$P_1 = \int_0^H 20z \, dz = \tfrac{1}{2} \times 20 \times 5^2 = 250 \text{ kN}$$

Example 18.2: Drained bearing capacity of a foundation Figure 18.24(a) shows a long foundation, 3 m wide and carrying a load F per metre out of the page, buried 1 m below the ground surface in dry soil, which has a friction angle $\phi' = 20°$. The bearing pressure (i.e. the total stress on the underside of the foundation) is $q = F/B$. For simplicity the soil is assumed to be weightless ($\gamma = 0$) except above foundation level, where $\gamma = 20 \text{ kN/m}^3$, so that the 1 m deep layer applied a uniform surcharge $\gamma D = 20 \text{ kPa}$ at foundation level. The idealized loads and stresses are shown in Fig. 18.24(b).

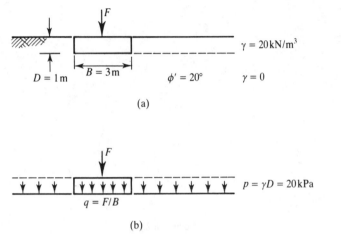

(a)

(b)   **Figure 18.24**

(a) Upper bound. Figure 18.25(a) shows a mechanism consisting of two wedges and Fig. 18.25(b) is the corresponding displacement diagram. For $\phi' = 20°$ suitable angles for the slip surfaces are $45° \pm 10°$ and all displacements are at angles $\psi = \phi' = 20°$ to the slip surfaces. (Notice that if $\phi' > 30°$ the directions ob and ab diverge and the mechanism is not compatible.)
From the geometry of Fig. 18.25(a),

$$L = B \tan^2(45° + \tfrac{1}{2}\phi') = 3 \tan^2 55° = 6.1 \text{ m}$$

and so the force applied by the stress p is

$$P = pL = 20 \times 6.1 = 122 \text{ kN}$$

From the geometry of Fig. 18.25(b), taking $\delta w_f = 1$,

$$\delta w_p = \tan(45° + \tfrac{1}{2}\phi') \tan(45° + \tfrac{3}{2}\phi') = \tan 55° \tan 75° = 5.3$$

For drained loading, from Eq. (18.12), $\delta W = 0$. The work done by the external loads is given by Eq. (18.9) with $\gamma = \gamma_w = 0$ for weightless and dry soil:

$$\delta E = F_u \, \delta w_f - P \, \delta w_p = (F_u \times 1) - (122 \times 5.3)$$

Hence, equating $\delta E = \delta W$,

$$F_u = 122 \times 5.3 = 647 \text{ kN}$$

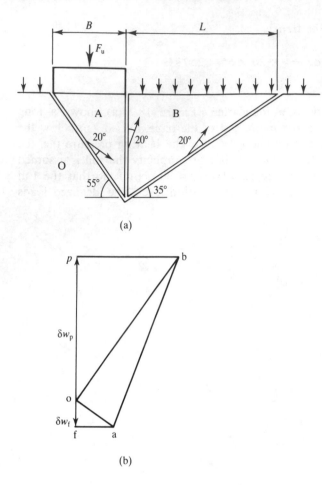

(a)

(b)

Figure 18.25

(b) Lower bound. Figure 18.26(a) shows a simple equilibrium stress field, symmetric about the centre-line, with two discontinuities. Figure 18.26(b) shows two Mohr circles of effective stress for the two regions of uniform stress below and to the side of the foundation. From the geometry of Fig. 18.26(b),

$$\frac{q_1}{\sigma'_h} = \frac{\sigma'_h}{p'} = \tan^2(45° + \tfrac{1}{2}\phi')$$

Hence,

$$F_1 = Bp' \tan^4(45° + \tfrac{1}{2}\phi') = 3 \times 20 \times \tan^4 55° = 250 \text{ kN}$$

Example 18.3: Drained bearing capacity of a foundation Better bound solutions for the foundation in Fig. 18.24 can be found using a mechanism which includes a slip fan and a stress field which includes a stress fan.

(a) Upper bound. Figure 18.27(a) shows a mechanism consisting of two wedges and a logarithmic spiral slip fan and Fig. 18.27(b) is the corresponding displacement diagram. For $\phi' = 20°$ the angles in the mechanism and in the displacement diagram are $45 \pm \tfrac{1}{2}\phi' = 55°$ or $35°$.

$q_l = F_l/B$

$p' = \gamma D$

q'_l p'

$\sigma'_h \rightarrow$ $\leftarrow \sigma'_h$

(a)

τ'

ϕ'

p' σ_h q'_l σ'

(b)

Figure 18.26

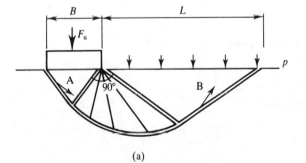

B L

F_u

p

A $90°$ B

(a)

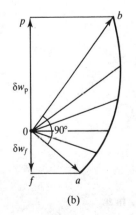

p b

δw_p

0 $90°$

δw_f

f a

(b)

Figure 18.27

From the geometry of Fig. 18.27(a), and making use of Eq. (18.6) with $\psi = \phi'$,

$$L = B \tan(45° + \tfrac{1}{2}\phi') \exp(\Delta\theta \tan \phi')$$

$$L = 3 \tan 55° \exp(\pi/2 \tan 20°) = 7.6 \text{ m}$$

and

$$P = pL = 20 \times 7.6 = 152 \text{ kN}$$

From the geometry of Fig. 18.27(b), taking $\delta w_f = 1$,

$$\delta w_p = \tan(45° + \tfrac{1}{2}\phi') \exp(\Delta\theta \tan \phi')$$

$$\delta w_p = \tan 55° \exp(\pi/2 \tan 20°) = 2.53$$

For drained loading, from Eq. (18.12), $\delta W = 0$. The work done by the external loads is given by Eq. (18.9) with $\gamma = \gamma_w = 0$ for weightless and dry soil:

$$\delta E = F_u \, \delta w_f - P \, \delta W_p$$

Equating $\delta E = \delta W$,

$$F_u = 152 \times 2.53 = 385 \text{ kN}$$

(b) Lower bound. Figure 18.28(a) shows an equilibrium stress field consisting of a region B where $\sigma'_1 \ (= q_1)$ is vertical and a region A where $\sigma'_3 \ (= p' = 20 \text{ kPa})$ is vertical. These are separated by a fan zone with a fan angle of 90°. There could be a similar stress field at the left-hand edge of the foundation. Figure 18.28(b) shows the two Mohr circles of effective stress for the two regions of uniform stress.

From Eq. (18.39),

$$s'_b = s'_a \exp(2\theta_f \tan \phi') = s'_a \exp(\pi \tan 20°) = 3.14 s'_a$$

(a)

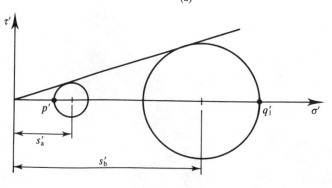

(b)

Figure 18.28

From the geometry of the Mohr circles,

$$s_a' = p'\left(\frac{1}{1 - \sin \phi'}\right) = 20\left(\frac{1}{1 - \sin 20°}\right) = \frac{20}{0.66}$$

$$s_b' = q_1'\left(\frac{1}{1 + \sin \phi'}\right) = q_1'\left(\frac{1}{1 + \sin 20°}\right) = q_1'\frac{1}{1.34}$$

Hence,

$$q_1' = 1.34 \times 3.14 s_a' = 1.34 \times 3.14 \times \frac{20}{0.66} = 128\text{kPa}$$

and

$$F_1 = q_1'A = 128 \times 3 = 384 \text{ kN}$$

Notice that these last upper and lower bound solutions are the same; this is because the mechanism in Fig. 18.27 corresponds to the stress field in Fig. 18.28.

FURTHER READING

Atkinson, J. H. (1981) *Foundations and Slopes*, McGraw-Hill, London.

Calladine, C. R. (1969) *Engineering Plasticity*, Pergamon Press, London.

Chen, W. F. (1975) *Limit Analysis and Soil Plasticity*, Elsevier, New York.

Sokolovskii, V. V. (1965) *Statics of Granular Media*, Pergamon Press, Oxford.

19.1 THEORY OF THE LIMIT EQUILIBRIUM METHOD

The limit equilibrium method is by far the most commonly used analysis for the stability of geotechnical structures. The steps in calculating a limit equilibrium solution are as follows:

1. Draw an arbitrary collapse mechanism of slip surfaces; this may consist of any combination of straight lines or curves arranged to give a mechanism.
2. Calculate the statical equilibrium of the components of the mechanism by resolving forces or moments and hence calculate the strength mobilized in the soil or the external forces (whichever is unknown).
3. Examine the statical equilibrium of other mechanisms and so find the critical mechanism for which the loading is the limit equilibrium load.

Remember that, as always, we must distinguish between cases of undrained and drained loading. For undrained loading the ultimate strength of the soil is given by

$$\tau = s_u \tag{19.1}$$

where s_u is the undrained shear strength. For drained loading where pore pressures can be determined from hydrostatic groundwater conditions or from a steady state seepage flownet, the strength is given by

$$\tau' = \sigma' \tan \phi'_c = (\sigma - u) \tan \phi'_c \tag{19.2}$$

where ϕ'_c is the critical state friction angle. These strengths give the ultimate collapse states and in order to design safe structures or to limit ground movements, they can be reduced by a factor F_s as described in later sections.

The limit equilibrium method combines features of the upper and lower bound methods. The geometry of the slip surfaces must form a mechanism that will allow collapse to occur, but they may be any shape so they need not meet all the requirements of compatibility (see Sec. 18.3). The overall conditions of equilibrium of forces on blocks within the mechanism must be satisfied, but the local states of stress within the blocks are not investigated. Although there is no formal proof that the limit equilibrium method leads to correct solutions, experience has shown that the method usually gives solutions that agree quite well with observations of the collapse of real structures and the method is firmly established among the techniques of geotechnical engineering.

19.2 SIMPLE LIMIT EQUILIBRIUM SOLUTIONS

Two simple problems, one for drained loading and one for undrained loading, are shown in Figs 19.1 and 19.2. These illustrate the general principles of the limit equilibrium method. Figure 19.1(a) shows part of a very long slope in soil where the pore pressures are zero. The problem is to determine the critical slope angle i_c when the slope fails. A mechanism could be a straight slip surface at a depth z as shown, and the forces on the block with length L down the surface are marked on the diagram. If the slope is very long, F_1 and F_2 are equal and opposite. The normal and shear forces on the slip surface are $T' = \tau' L$ and $N' = \sigma' L$ and the weight is $W = \gamma L z \cos i$. Figure 19.1(b) is a polygon of these forces which closes (i.e. the forces are in equilibrium) when

$$\frac{T'}{N'} = \frac{\tau'}{\sigma'} = \tan i_c \qquad (19.3)$$

Hence, from Eq. (19.2), the limit equilibrium solution is

$$i_c = \phi'_c \qquad (19.4)$$

(a)   (b)

Figure 19.1 Limit equilibrium solution for stability of an infinite slope for drained loading.

Figure 19.2 Limit equilibrium solution for the bearing capacity of a foundation for undrained loading.

Figure 19.3 Limit equilibrium solution for the bearing capacity of a foundation for undrained loading.

Strictly we should consider other possible mechanisms with combinations of curved and straight slip surfaces, but it is fairly obvious that the mechanism illustrated in Fig. 19.1 is one of the most critical. The solution $i_c = \phi'_c$ can also be obtained as an upper bound and as a lower bound so it is an exact solution.

Figure 19.2(a) shows a section of a foundation with width B and unit length out of the page so that the width B is equal to the foundation area A. The foundation is loaded undrained and the undrained strength of the soil is s_u. The problem is to determine the collapse load F_c or the ultimate bearing capacity $q_c = F_c/A$. A mechanism could be a circular slip surface with centre O at the edge of the foundation. The rotating block of soil is in equilibrium when the moments about O balance and

$$F_c \times \tfrac{1}{2}B = s_u B \,\widehat{ST} \tag{19.5}$$

where $\widehat{ST} = \pi B$ is the length of the arc ST. Notice that the lines of action of the weight W of the soil block and the normal stresses on the circular slip surfaces act through O and so their moments about O are zero. From Eq. (19.5) we have

$$F_c = 2\pi B s_u \tag{19.6}$$

As before we should now consider other possible mechanisms with combinations of straight and curved slip surfaces to seek the minimum value of F_c, which will be the limit equilibrium solution. Figure 19.3 shows a circular slip surface with its centre at a height h above the ground surface. Readers should show that the minimum value for this mechanism is $F_c = 5.5Bs_u$ when $h/B = 0.58$; one way to do this is to take trial values of h and plot F_c against h to determine the minimum value of F_c.

Remember that in Chapter 18 we obtained equal upper and lower bound solutions (i.e. an exact solution) for a foundation on undrained soil as $F_c = (2 + \pi)Bs_u$ (see Eq. 18.46) and so, in this case, the best limit equilibrium solution with a circular arc slip surface overestimates the true solution by less than 10 per cent.

19.3 COULOMB WEDGE ANALYSIS

Calculation of the loads required to maintain the stability of a retaining wall provides a convenient example to illustrate both the basic features of the limit equilibrium method and a number of special features of the method. Solutions are particularly simple as a mechanism can be constructed from a single straight slip surface. This calculation was first developed by Coulomb in about 1770 and is one of the earliest engineering calculation still in current use, although with a number of modifications.

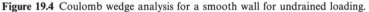

Figure 19.4 Coulomb wedge analysis for a smooth wall for undrained loading.

Figure 19.4(a) shows a section of a smooth wall with unit length out of the page supporting soil that is undrained. The horizontal force on the wall necessary to prevent the soil collapsing into the excavation is P_a and this is called the active force (see Sec. 21.1). (In practice, vertical cracks may form in the ground near the top of the wall; I will consider the influence of tension cracks later, but for the present I will assume that they do not occur.) A mechanism can be constructed from a single straight slip surface at an angle α and there must be slip surfaces between the soil and the wall as shown.

The forces acting on the triangular wedge are shown in Fig. 19.4(b). There is no shear force between the soil and the smooth wall. The directions of all the forces are known and the magnitudes of P_a and N are unknown; the magnitude of the shear force T is given by

$$T = s_u L \tag{19.7}$$

where s_u is the undrained strength and L is the length of the slip surface; T acts up the surface as the wedge moves down into the excavation. With two unknowns the problem is statically determinate and a solution can be found by resolution of the forces; notice that if you resolve in the direction of the slip surface N does not appear and P_a can be found directly. Alternatively, the solution can be found graphically by constructing the closed polygon of forces in Fig. 19.4(c).

To obtain the limit equilibrium solution you must vary the angle α to find the maximum, or critical, value for P_a. If you do this you will find that the critical angle is $\alpha = 45°$ and the limit equilibrium solution is

$$P_a = \tfrac{1}{2}\gamma H^2 - 2s_u H \tag{19.8}$$

Notice that if we put $P_a = 0$ we obtain

$$H_c = \frac{4s_u}{\gamma} \tag{19.9}$$

which is a limit equilibrium solution for the undrained stability of an unsupported trench.

This analysis can be extended quite simply to include the effects of foundation loads, water in the excavation and shear stresses between the soil and a rough wall. The additional forces are

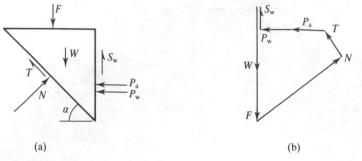

Figure 19.5 Coulomb wedge analysis for a rough wall for undrained loading.

(a)   (b)

shown in Fig. 19.5(a) and the corresponding polygon of forces is shown in Fig. 19.5(b). The shear force on the wall S_w is given by

$$S_w = s_w H \tag{19.10}$$

where s_w is the ultimate shear stress between the soil and the wall; obviously s_u must be in the range $0 \leqslant s_w \leqslant s_u$ depending on the roughness of the wall. Free water in the excavation applies a total force P_w to the wall, given by

$$P_w = \tfrac{1}{2}\gamma_w H_w^2 \tag{19.11}$$

where H_w is the depth of water in the excavation. For the undrained case the pore pressures in the soil do not come into the calculation and will not be in equilibrium with the water pressures in the excavation. Again the only unknowns are the magnitudes of the forces N and P_a, so the problem is statically determinate. The limit equilibrium solution is the maximum value of P_a and coincides with the critical slip surface.

The case shown in Fig. 19.6 is similar to that in Fig. 19.4 except that the soil is drained and dry so pore pressures are zero. The forces on the triangular wedge are shown in Fig. 19.6(a). There are now three unknown forces, T', N' and P_a', but the forces T' and N' are related by Eq. (19.2) so the resultant of T' and N', shown by the broken line, is at an angle ϕ_c' to the direction of N'. (The primes are added to these forces because they are associated with the effective stresses in the dry soil.) This now provides sufficient information to construct the force polygon shown in Fig. 19.6(b) to calculate the magnitude of P_a. To obtain the limit equilibrium solution you must vary the angle α to find the critical value for P_a. This occurs when $\alpha = 45° + \tfrac{1}{2}\phi_c'$, and the limit equilibrium solution is

$$P_a = \tfrac{1}{2}\gamma H^2 \tan^2(45 - \tfrac{1}{2}\phi_c') \tag{19.12}$$

(a)   (b)

Figure 19.6 Coulomb wedge analysis for a smooth wall for drained loading.

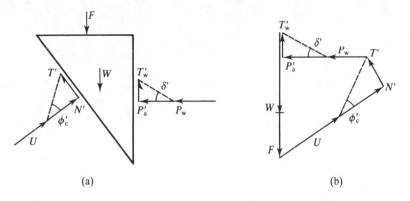

(a)   (b)

Figure 19.7 Coulomb wedge analysis for a rough wall for drained loading.

This solution was developed by Rankine in about 1850 (but in a different way) and is really a case of the Coulomb wedge analysis.

Again the analysis can be extended to include external loads, water in the excavation, pore pressures and shear stresses between the soil and a rough wall. The additional forces are shown in Fig. 19.7(a) and the corresponding polygon of forces is shown in Fig. 19.7(b). For simplicity the water table is assumed to be the same in the soil and in the excavation, so there is no seepage; later I will examine the case where the excavation is dewatered and there is a steady state seepage flownet in the soil. The force U is the sum (or integral) of the pore pressures over the slip surface and is found by summing Eq. (6.4) over the length L of the slip surface. The shear force T' is given by

$$T'' = N' \tan \phi'_c = (N - U) \tan \phi'_c \tag{19.13}$$

Similarly, the shear force between the soil and the wall is given by

$$T'_w = P'_a \tan \delta'_c \tag{19.14}$$

where δ'_c is the ultimate or critical friction angle between the soil and the wall; obviously δ'_c must be in the range $0 \leqslant \delta'_c \leqslant \phi'_c$ depending on the roughness of the wall. Notice that the total normal force on the vertical face of the soil is $P'_a + P_w$ (i.e. the sum of the force from the support prop and the force from the free water).

In Figs 19.4 and 19.6 the major principal planes are horizontal because the shear stress on the wall is zero and $\sigma_z > \sigma_h$. In Sec. 2.6 we found that zero extension lines (i.e. lines of zero strain) were at angles $\alpha = 45° + \frac{1}{2}\psi$ to the major principal plane and planes where the stress ratio was $\tau'/\sigma' = \tan \rho'$ were at angles $\alpha = 45° + \frac{1}{2}\rho'$ to the major principal plane. For undrained loading $\psi = 0$ and for drained loading, at the critical state $\rho' = \phi'_c$. Hence the critical surfaces in these limit equilibrium solutions coincide with the critical planes and zero extension lines obtained from the Mohr circle constructions discussed in Chapter 2. In Figs 19.5 and 19.7 there are shear stresses between the wall and the soil, so horizontal and vertical planes are not principal planes and the critical surfaces are not necessarily at angles $\alpha = 45°$ or $45° + \frac{1}{2}\phi'_c$ to the horizontal.

19.4 SIMPLE SLIP CIRCLE ANALYSES FOR UNDRAINED LOADING

A mechanism in which the slip surface is a circular arc—or a slip circle—as shown in Fig. 19.2, is very commonly used in routine limit equilibrium analyses in geotechnical engineering. The

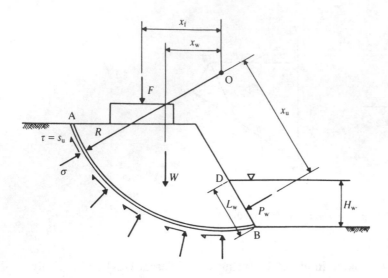

Figure 19.8 Slip circle method for undrained loading.

methods of solution are different for drained and for undrained loading and we will consider each separately.

Figure 19.8 shows a section of a slope with a foundation at the top and water in a river or lake at the toe. There is a mechanism consisting of a single circular arc with centre at O. The forces on the mechanism are due to the foundation load F, the weight of the soil W, the free water P_w and the shear stresses in the soil $T = s_u \widehat{AB}$ where $\widehat{AB}$ is the length of the arc AB; these forces have lever arms x and R as shown. Taking moments about O, the foundation and slope are just stable when

$$Wx_w + Fx_f - P_w x_u = s_u \widehat{AB} R \tag{19.15}$$

The limit equilibrium solution must be found by searching for the critical slip circle by varying the radius and the position of the centre. Notice in Fig. 19.8 that the normal stresses on the slip circle are radial and pass through the origin, so they have no moment about O. Calculation of values for Wx_w and $s_u \widehat{AB} R$ can be simplified by dividing the mechanism into a number of vertical slices and tabulating the results as in Fig. 19.18 in Example 19.3.

19.5 SLIP CIRCLE METHOD FOR DRAINED LOADING—THE METHOD OF SLICES

Figure 19.9(a) shows a slope with a steady state seepage flownet to a drain at the toe. Pore pressures anywhere in the slope can be calculated from the flownet, as described in Chapter 17. Figure 19.9(b) shows a slip circle mechanism and, taking moments about the centre O, the slope is just stable when

$$Wx = R \int_{AB} \tau' \, dl \tag{19.16}$$

where the shear stresses are given by

$$\tau' = (\sigma - u) \tan \phi'_c \tag{19.17}$$

Although we can calculate values for the pore pressure u around the slip circles we cannot, at

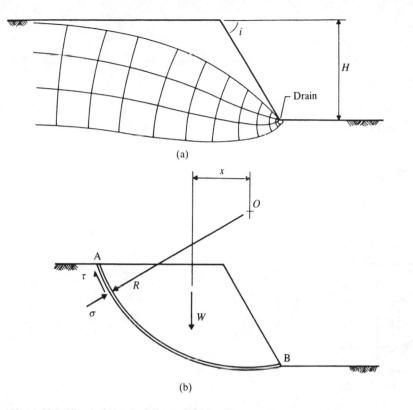

(a)

(b)

Figure 19.9 Slip circle method for drained loading.

present, calculate the normal stresses σ. Thus the simple analysis which served for undrained loading for which $\tau = s_u$ cannot be used for drained loading.

The approach adopted for the method of slices is to subdivide the mechanism into a number of approximately equal vertical slices and examine the statical equilibrium of the slices and, by summation, of the whole mechanism. Figure 19.10(a) shows the mechanism of Fig. 19.9 divided into four slices, of which a typical slice FGHJ is shown in Fig. 19.10(b). The total forces on the slice shown in Fig. 19.10(b) are its weight W, and total normal and shear forces N and T on the base FJ, and forces F_1 and F_2 from adjacent slices. The interslice forces F_1 and F_2 are not necessarily equal and opposite, and their resultant F acts at a height a above the centre of the base of the slice and at an angle θ to the horizontal. The total normal and shear forces on the base of the slice are related by

$$T = (N - U) \tan \phi'_c \tag{19.18}$$

where the forces $T = \tau l$, $N = \sigma l$ and $U = ul$, where l is the length of the base FJ. Summing for all the slices gives

$$\sum T = \sum (N - U) \tan \phi'_c \tag{19.19}$$

The interslice forces such as F may be decomposed into horizontal and vertical components E and X. In the slip circle method the boundaries between adjacent slices are not slip surfaces and so nothing can be said at present about the magnitude, direction or point of application of the force F in Fig. 19.10. Considering the forces on the block FGHJ in Fig. 19.10(b), the magnitudes, direction and points of application are known for W and U, the directions and

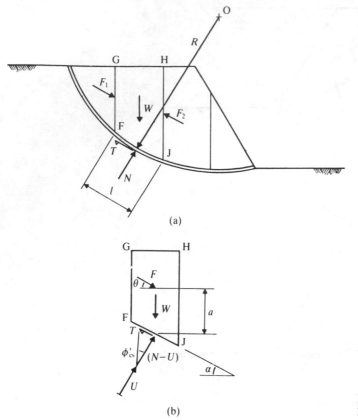

(a)

(b)

Figure 19.10 Slip circle method for drained loading—method of slices.

points of application are known for T and N, but nothing is known about the force F. Thus there are five unknowns: T, N, F, a and θ. We can obtain three equations by resolution of forces and by taking moments following the usual rules of statics. These, together with Eq. (19.18), lead to a possible total of four equations and each slice is statically indeterminate. To obtain a solution for the method of slices for drained loading we are obliged to make at least one simplifying assumption in order to make the problem statically determinate. There are a number of such solutions, each based on a different simplifying assumption. For the present I will consider the two commonest of these solutions.

(a) The Swedish Method of Slices (Fellenius, 1927)

Here it is assumed that the resultant F of the interslice forces is zero for each slice and thus F, a and θ vanish. Each slice is then statically determinate, and from Fig. 19.11 we have

$$T = W \sin \alpha \qquad N = W \cos \alpha \tag{19.20}$$

where α is the average inclination of the slip surface at the base of the slice. Here we may calculate T and N for each slice and, for equilibrium, making use of Eq. (19.19),

$$\sum W \sin \alpha = \sum (W \cos \alpha - ul) \tan \phi_c' \tag{19.21}$$

where u is the average pore pressure over the length l of the base of each slice. Instead of making use of Eqs (19.20) we may calculate T and N for each slice from force polygons like those shown in Fig. 19.10(b). The calculations are assisted by the use of a table such as that shown in Fig.

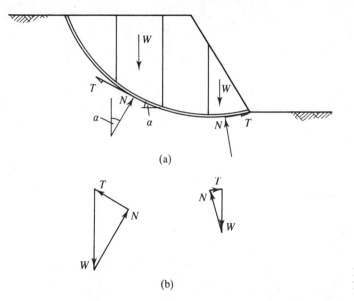

(a)

(b)

Figure 19.11 Slip circle method for drained loading—Swedish method.

19.19(c) in Example 19.4. As before, it is necessary to examine a number of different mechanisms to locate the critical slip circle; the slope is taken to be in a state of collapse if Eq. (19.21) is satisfied for any mechanism.

(b) The Bishop Routine Method (Bishop, 1955)

Here it is assumed that the resultant of the interslice forces is horizontal. Hence $\theta = 0$ as shown in Fig. 19.12 and each slice is statically determinate. After resolving, taking moments and summing over the whole mechanism, the solution comes out in the form

$$\sum W \sin \alpha = \sum \frac{(W - ub) \sec \alpha \tan \phi'_c}{1 + \tan \alpha \tan \phi'_c} \tag{19.22}$$

where b is the width of each slice. In practice, evaluation of Eq. (19.22) is simplified if use is made of a table similar to that in Fig. 19.19(c). As before, it is necessary to examine a number of different mechanisms to locate the critical slip circle; the slope is then taken to be in a state of collapse if Eq. (19.22) is satisfied for any mechanism.

19.6 OTHER LIMIT EQUILIBRIUM METHODS

So far we have considered mechanisms consisting either of a single straight slip surface or a circular arc. The limit equilibrium method is not restricted to these geometries and there are two other commonly used arrangements of slip surfaces.

Figure 19.13 shows a mechanism consisting of several straight slip surfaces forming two triangular wedges and a block; this mechanism is appropriate where a layer of relatively weak soil occurs within the slope as shown. The shear and normal forces across each slip surface are marked. In this case, unlike the method of slices, the soil in the vertical slip surfaces is at failure and so the shear stresses can be determined from either Eq. (19.1) or (19.2) and the lengths of the slip surfaces. Working from the left-hand wedge towards the right, the forces on each block are statically determinate.

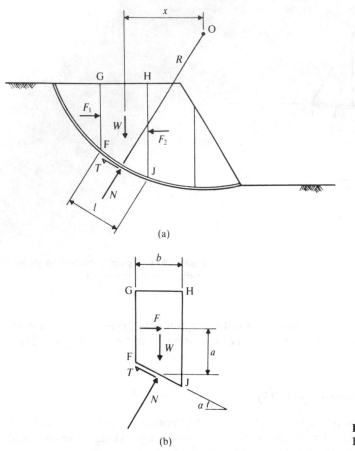

(a)

(b)

Figure 19.12 Slip circle for drained loading—Bishop's method.

Figure 19.14 shows a mechanism in which there is a single continuous slip surface of general shape. The solution is found using the method of slices, as described above, for which at least one simplifying assumption is required. Thus the Swedish method (X and $E = 0$) or the Bishop routine method ($X = 0$) can be applied to general slip surfaces. Other solutions were developed by Janbu (1973) and by Morgenstern and Price (1965). You can see that all these named methods (Swedish, Bishop, Janbu, Morgenstern and Price, and others) are basically limit equilibrium solutions using the method of slices with different assumptions to avoid the problem of statical indeterminacy.

Weak soil

Figure 19.13 Wedge method.

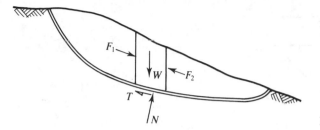

Figure 19.14 General slip surface method.

19.7 LIMIT EQUILIBRIUM SOLUTIONS

Although the limit equilibrium method is approximate and requires a number of basic assumptions it has advantages over other methods. It is quite general and can be applied to walls, slopes or foundations, or to any combination of these. The method can be adapted for cases where the soil has layers with different properties or irregularly shaped boundaries.

The calculations for determining the forces on slices and for varying the geometry of the mechanism of slip surfaces are largely repetitive and there are a number of computer programs for the stability of geotechnical structures that make use of the limit equilibrium method.

19.8 SUMMARY

1. The basic limit equilibrium method requires that blocks of soil inside a mechanism of slip surfaces are in statical equilibrium.
2. Mechanisms consist of slip surfaces which may be straight lines, arcs of circles (in the slip circle method) or any general shape.
3. Coulomb and Rankine analyses apply for mechanisms consisting of a single straight slip surface and the equilibrium calculations can be done using polygons of forces.
4. For undrained analyses with slip circles solutions can be found by taking moments about the centre of the circle.
5. For drained analyses with slip circles or with any general slip surface the problem is statically indeterminate and solutions are found using the method of slices with one of a number of alternative assumptions.

WORKED EXAMPLES

Example 19.1: Coulomb wedge analysis for undrained loading The trench shown in Fig. 19.15 is supported by rough sheet piles held apart by struts, 1 m apart out of the page, placed so that the piles do not rotate. The trench is part filled with water as shown.

$q = 80\,\text{kPa}$

$s_u = 40\,\text{kPa} \quad \phi_c' = 25°$
$s_w = 20\,\text{kPa} \quad \delta_c' = 15°$
$\gamma = 20\,\text{kN/m}^3$

4 m

5 m

Figure 19.15

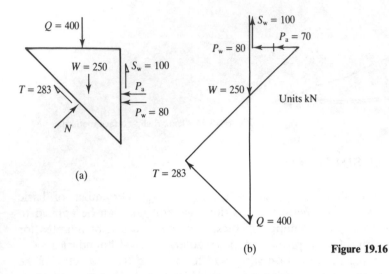

Figure 19.16

For undrained loading a suitable Coulomb wedge is formed by a single slip plane at 45° to the horizontal and Fig. 19.16(a) shows the forces on the wedge. The magnitudes of the known forces are

$$Q = qH = 80 \times 5 = 400 \text{ kN}$$

$$W = \tfrac{1}{2}\gamma H^2 = \tfrac{1}{2} \times 20 \times 5^2 = 250 \text{ kN}$$

$$T = \sqrt{2}Hs_u = \sqrt{2} \times 5 \times 40 = 283 \text{ kN}$$

$$S_w = s_w H = 20 \times 5 = 100 \text{ kN}$$

$$P_w = \tfrac{1}{2}\gamma_w H_w^2 = \tfrac{1}{2} \times 10 \times 4^2 = 80 \text{ kN}$$

The force polygon is shown in Fig. 19.16(b). Scaling from the diagram, or by calculation,

$$P_a = 70 \text{ kN}$$

Example 19.2: Coulomb wedge analysis for drained loading For drained loading a suitable Coulomb wedge is formed by a slip plane at $\alpha = 45° + \tfrac{1}{2}\phi_c' = 57\tfrac{1}{2}°$ to the horizontal and Fig. 19.17(a) shows the forces on the wedge. The magnitudes of the known forces are

$$Q = qH \tan(90° - \alpha) = 80 \times 5 \times \tan 32.5° = 255 \text{ kN}$$

$$W = \tfrac{1}{2}\gamma H^2 \tan(90° - \alpha) = \tfrac{1}{2} \times 20 \times 5^2 \times \tan 32.5° = 159 \text{ kN}$$

$$P_w = \tfrac{1}{2}\gamma_w H_w^2 = \tfrac{1}{2} \times 10 \times 4^2 = 80 \text{ kN}$$

$$U = \tfrac{1}{2}\gamma_w H_w^2 \times \frac{1}{\sin \alpha} = \tfrac{1}{2} \times 10 \times 4^2 \times \frac{1}{\sin 57.5°} = 95 \text{ kN}$$

and the other information is

$$T' = (N - U) \tan 25°$$

$$T_w' = P' \tan 15°$$

The force polygon is shown in Fig. 19.17(b). Scaling from the diagram or by calculation,

$$P_a' = 245 \text{ kN}$$

(a)　　　　　　　　　　　(b)　　　　**Figure 19.17**

Example 19.3: Undrained slope stability　Figure 19.18(a) shows a slope and a slip circle divided into slices. For the case where the soil is undrained, replacing s_u with s_u/F_s and making use of Eq. (19.15),

$$F_s = \frac{\sum s_u R l}{\sum W x}$$

(a)

Slice	Area (m^2)	W (kN)	x (m)	Wx (kNm)	l (m)	$s_u R l$ (kNm)
1	3.8	76	9.2	699	4.0	1920
2	14.4	288	6.8	1958	3.7	1776
3	15.3	306	3.8	1163	3.2	1536
4	13.8	276	0.8	221	3.0	1440
5	10.2	204	−2.2	−449	3.0	1440
6	3.8	75	−4.7	−353	3.2	1536
Totals				3239		9648

(b)　　　　　　　　　　　　　**Figure 19.18**

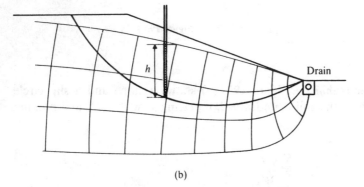

(a)

(b)

Slice	Area (m^2)	W (kN)	α	$W \sin \alpha$ (kN)	$W \cos \alpha$ (kN)	h (m)	u (kN/m^2)	l (m)	ul (kN)	$W \cos \alpha - ul$ (kN)
1	3.8	76	54°	61	45	1.4	13.7	4.0	55	−10
2	14.4	288	36°	169	233	3.0	29.4	3.7	109	124
3	15.3	306	20°	105	288	4.0	39.2	3.2	126	162
4	13.8	276	4°	19	275	3.8	37.3	3.0	112	163
5	10.2	204	−10°	−35	201	2.8	27.5	3.0	82	119
6	3.8	76	−25°	−32	69	1.4	13.7	3.2	44	25
Totals				287						583

(c)

Figure 19.19

The table in Fig. 19.18(b) gives the calculations for each slice and, summing over the whole mechanism,

$$F_s = \frac{9648}{3238} = 2.98$$

You should now repeat the calculations with different values of the radius R and different positions for the centre O to find the lowest value of F_s.

Example 19.4: Drained slope stability   Figure 19.19(a) shows a slope and a slip circle divided into slices and Fig. 19.19(b) shows a flownet sketched for steady state seepage towards a drain at the toe of the slope. The pore pressure at any point on the slip circle can be estimated from the height of the water in a standpipe on an equipotential as shown.

Replacing $\tan \phi_c'$ with $\tan \phi_c'/F_s$ and making use of Eq. (19.21) for the Swedish method of slices,

$$F_s = \frac{\sum (W \cos \alpha - ul)}{\sum W \sin \alpha} \tan \phi_c'$$

The table in Fig. 19.19(c) gives the calculations for each slice and, summing over the whole mechanism,

$$F_s = \frac{583}{287} \tan 30° = 1.17$$

You should now repeat the calculations with different circles to find the lowest value of F_s. Notice that near the toe of the slope the seepage becomes approximately parallel to the surface (see Sec. 20.6) and there is the possibility of local instability, which should be investigated.

REFERENCES

Bishop, A. W. (1955) 'The use of the slip circle in the stability analysis of earth slopes', *Geotechnique*, **5**, 7–17.
Fellenius, W. (1927) *Erdstatische Berechnungen*, W. Ernst und Sohn, Berlin.
Janbu, N. (1973) 'Slope stability computations', in *Embankment Dam Engineering, Casagrande Memorial Volume*. Hirschfield and Poulos (eds), Wiley, New York.
Morgenstern, N. R. and V. E. Price (1965) 'The analysis of the stability of general slip surfaces', *Geotechnique*, **15**, 79–93.

FURTHER READING

Atkinson, J. H. (1981) *Foundations and Slopes*, McGraw-Hill, London.
Bromhead, E. N. (1986) *The Stability of Slopes*, Surrey University Press, London.
Heyman, J. (1972) *Coulomb's Memoir on Statics*, Cambridge University Press, Cambridge.

CHAPTER TWENTY

STABILITY OF SLOPES

20.1 INTRODUCTION

The surface of the earth is very rarely flat and so there are slopes nearly everywhere. Even relatively flat ground often has rivers and drainage channels with side slopes. Slopes may be natural, due to erosion by rivers or the sea, or man-made by excavation or fill. Man-made slopes for roads and dams are permanent, but temporary slopes are required during construction of foundations and underground structures.

The geometry of a slope may be characterized by its angle i and height H, as shown in Fig. 20.1. The loads on the slope are due to the self-weight of the soil and to external loads, which may come from foundations at the top or water in the excavation. A special case of a slope is a vertical cut, such as the sides of a trench, where $i = 90°$. In the soil behind any slope there will be shear stresses and these are required to maintain the slope. Materials that cannot sustain shear stresses cannot have slopes, so water in a glass has a level surface.

During excavation of a slope the mean normal total stresses will be decreased due to removal of soil from the excavation, while during construction of an embankment the mean normal total stresses will increase as more fill is placed. In both cases, however, the shear stresses increase as the height and/or slope angle increase. I will call any kind of slope construction *loading* because the shear stresses increase irrespective of what happens to the mean normal total stress.

If a slope is too steep or too high it will fail and there will be a slip or landslide, as illustrated in Fig. 20.2. The slip will stop when the height and angle are critical (H_c and i_c) and the slope has a factor of safety of unity. Rock slopes can be very steep and very high (look at a photograph of Everest, for example), but soil slopes are much more modest, with angles from 10° to 30°

Soil (γ, s_u or ϕ')
Water pressures u

H

i

Figure 20.1 Geometry of a simple slope.

Figure 20.2 Simple slope failure.

and, for steeper angles, heights up to 20 m. The best laboratory to study slope stability is at the seaside where you should dig a hole in the beach and construct a sandcastle.

20.2 TYPES OF INSTABILITY

Slope instabilities involve large ground movements and usually require a mechanism of slip surfaces. Mechanisms can have a number of different configurations and some typical ones are illustrated in Fig. 20.3. In Fig. 20.3(a) and (b) the soil is homogeneous and the position of the slip surface (deep or shallow) is governed largely by the pore pressures. In Fig. 20.3(c) and (d) the geometry of the slip surface is controlled by strong or weak layers. Figure 20.3(e) illustrates a mud flow where there are very large homogeneous strains. Figure 20.3(f) illustrates a block

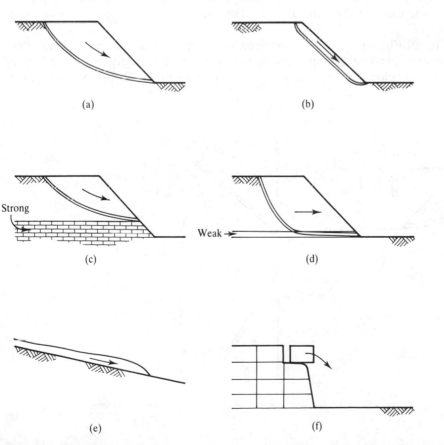

Figure 20.3 Types of slope failure.

failure of a fissured or jointed soil; this mechanism is not compatible because a vertical crack has opened.

Figure 20.3 illustrates only a few characteristic slope instability mechanisms and there are others. Many real landslides and slope failures involve combinations of several different mechanisms and can be quite complicated.

20.3 STRESS CHANGES IN SLOPES

Natural slopes are usually eroded very slowly and the soil is essentially drained so that pore pressures are governed by steady state seepage from the ground towards the excavation. Man-made slopes are often constructed quite quickly and in clays the soil will be essentially undrained during the excavation.

The changes of total and effective stress during undrained slope excavation are illustrated in Fig. 20.4. In Fig. 20.4(a) the total stresses on a typical element on a slip surface are τ and σ and the pore pressure is illustrated by the rise of water in a standpipe. (For simplicity the excavation is kept full of water so that the phreatic surface is level and the initial and final pore pressures are the same.)

In Fig. 20.4(b) the total stress path is A $\rightarrow$ B; this corresponds to a reduction in σ due to the excavation and an increase in τ because the slope height and/or angle are increased. The effective stress path is A$'$ $\rightarrow$ B$'$, which corresponds to undrained loading at constant water content, as shown in Fig. 20.4(c). The exact effective stress path A$'$ $\rightarrow$ B$'$ in Fig. 20.4(b) will depend on the characteristics of the soil and its initial state or overconsolidation ratio, as discussed in Chapter 11.

As shown in Fig. 20.4(b), the pore pressure immediately after construction u_i is less than the steady pore pressure u_0 and so the initial excess pore pressure $\bar{u}_i$ is negative (i.e. the level of water in the standpipe is below the phreatic surface, as shown in Fig. 20.4(a). As time passes

Figure 20.4 Stress and pore pressure changes in a stable slope.

the total stresses remain unchanged at B (because the geometry of the slope remains the same) but the negative excess pore pressures dissipate and rise. The effective stress path is B′ → C′ and this corresponds to swelling and a reduction in mean normal effective stress, as shown in Fig. 20.4(b) and (c). The final state at C′ corresponds to a steady state pore pressure after consolidation (swelling) u_c; in the example shown $u_c = u_0$ but the arguments would be the same if u_c was different from u_0, which would be the case if the excavation was drained of water.

The slope will fail if the states of all elements along the slip surface reach the critical state line: if B′ reaches the critical state line the slope fails during undrained excavation and if C′ reaches the critical state line the slope fails some time after construction. The distance of the effective stress points B′ or C′ from the critical state line is a measure of the factor of safety of the slope and Fig. 20.4 demonstrates that the factor of safety of a slope decreases with time.

This means that the critical time in the life of a slope is in the long term when the pore pressures have come into equilibrium with the steady state seepage flownet. Consequently, a permanent slope should be designed for the long-term, fully drained, condition. Temporary slopes that are required to stand for very short periods are often designed as undrained, but remember that just because a slope or a trench is standing now does not mean that it will still be stable in 10 minutes time. Slopes and excavations are very dangerous; many people are killed by trench failures which occur as the effective stresses move from B′ towards C′ in Fig. 20.4. In the design of temporary excavations the important question is not so much the undrained stability but how quickly the pore pressures will increase.

If a slope fails the total stresses change as the angle and height reduce as shown in Fig. 20.5(a). Figure 20.5(b) shows stress paths for a steep slope failing during undrained excavation. The effective stress path is A′ → B′ and this ends on the critical state line where the undrained

(a)

(b)

(c)

Figure 20.5 Stress and pore pressure changes in failing slopes.

strength is s_u. The total stress path would like to continue to X, corresponding to the initial slope angle i_x, but cannot; therefore the slope geometry changes and the mean slope angle i_c and height H_c correspond to total stresses at B. Figure 20.5(c) shows stress paths for a slope that fails some time after excavation. The state immediately after excavation is B and B' and failure occurs at C and C' when the pore pressure is u_f. Subsequently, as the pore pressures continue to rise, the effective stresses move along C' → D' down the critical state line and the total stresses move more or less along C → D due to unloading (i.e. reduction) of the shear stress as the slope angle decreases. The slope will reach a stable state when the pore pressure u_c is the final steady state pore pressure.

These analyses and the stress paths shown in Figs 20.4 and 20.5 are simplified and idealized but they illustrate the essential features of the behaviour of slopes during and after construction. Notice the critical importance of changing pore pressures with time and their influence on stability. The examples were for excavated slopes where pore pressures decreased during undrained excavation. In man-made compacted soils the initial pore pressures are negative because the fill is unsaturated and so the initial states at B and B' are more or less the same for cut and fill slopes.

20.4 INFLUENCE OF WATER ON STABILITY OF SLOPES

Water influences slope stability in several fundamentally different ways and these are illustrated by commonly observed failures. Firstly, slopes may fail well after completion of excavation due to dissipation of negative excess pore pressures and swelling and softening of the soil, as discussed in Sec. 20.3. Secondly, slopes in river banks, lakes and trenches may fail if the external water level is quickly lowered. Thirdly, slopes often fail after periods of heavy rainfall.

Free water in a river or lake, or in a water-filled trench, applies total stresses σ_w to a soil surface, as shown in Fig. 20.6. These total stresses help to support the slope which may fail if the support is removed. (In practice temporary excavations for piles and retaining walls are supported by a slurry of bentonite clay, or some other natural or artificial mud, with unit weight greater than that of water.) Notice that after undrained excavation the pore pressures in the soil may not be in equilibrium with the free water in the excavation.

Slope failures after rainfall, or after changes in the groundwater conditions, are due to increases in the pore pressures which lead to reductions in effective stress and the strength. (Notice that the soil remains saturated while pore pressures change and there is no question of the rainwater lubricating the soil—this is an entirely false interpretation.) In order to calculate the pore pressures in a slope under steady state conditions it is necessary to draw a flownet as described in Chapter 17. Figure 20.7 shows typical flownets for steady state seepage towards a drain in a road cutting and for flow parallel to a long slope. Pore pressures are given by the heights of water in standpipes, and this will be to the same level for all standpipes inserted to the same equipotential. In Fig. 20.7 the level of water in the standpipes inserted to the phreatic surface is at the phreatic surface and this gives the level for all standpipes at a particular

Figure 20.6 Loads on slopes from water in the excavation.

(a)

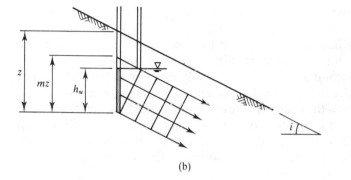

(b)

Figure 20.7 Flownets for steady state seepage in slopes.

equipotential. From the geometry of Fig. 20.7(b),

$$h_w = mz \cos^2 i \qquad (20.1)$$

while for the flownet in Fig. 20.7(a) pore pressures may be found by graphical constructions. For dry soil $m = 0$ and if the phreatic surface is at ground level $m = 1$.

20.5 CHOICE OF STRENGTH PARAMETERS AND FACTOR OF SAFETY

There are a number of possible criteria for defining soil strength (see Chapter 9). The most important of these are the distinction between the undrained strength s_u and the drained or effective stress strength and the distinction between the peak strength, the ultimate or critical state strength and the residual strength.

The choice between the undrained strength s_u and the drained strength is relatively simple and straightforward. For temporary slopes and cuts in fine-grained soils with low permeability choose the undrained strength s_u and carry out an analysis in total stresses. If you do this remember that the analysis is valid only so long as the soil is undrained and the stability will deteriorate with time as the pore pressures increase and the soil swells and softens.

For any permanent slope the critical conditions are at the end of swelling when pore pressures have reached equilibrium with a steady state seepage flownet or with hydrostatic conditions. In this case choose an effective stress strength and calculate the pore pressures separately. Analyses for slopes where the excess pore pressures due to excavation have only partially dissipated are beyond the scope of this book.

Further choices must be made between the peak, critical state and the residual strength

Figure 20.8 Variation of strength with displacement.

together with an appropriate factor of safety. The general variation of shearing resistance in soil with deformation or strain is illustrated in Fig. 20.8 (this is essentially the same as that shown in Fig. 9.3). The first thing to note is that if a major landslide or slope failure has occurred in a clay slope in the past the soil may have already reached its residual state. In this case new construction, either excavations or loading, may reactivate the old movements and the appropriate strength is the residual friction angle ϕ'_r. Detection of pre-existing landslides, some of which may be geologically very old, requires very detailed and careful ground investigations.

In the absence of pre-existing failures the choice is between the peak or the ultimate strength. In uncemented soils the peak strength is associated with dilation and occurs at relatively small strains or displacements of the order of 1 per cent or 1 mm. The ultimate (or critical state) strength is the shearing resistance for constant volume straining and occurs at strains or displacements of the order of 10 per cent or 10 mm.

Engineers designing slopes will not generally be concerned with ground movements (unlike designs for foundations and retaining walls) and they will simply want to ensure that the slope does not fail. In many slopes ground movements and strains are relatively large and exceed the small movements required to mobilize the peak state. If a steep slope fails it will come to rest when the geometry of the slope is in equilibrium with the ultimate or critical state strength and the pore pressures. There is ample evidence that the stability of cut and fill slopes is controlled by the critical state strength, with a factor of safety close to unity.

For the stability analyses of slopes described in the next sections I will choose the critical state strength parameters s_u or ϕ'_c. These strength parameters may be a little conservative in some cases but their use will lead to safe designs. If there is evidence of pre-existing landsliding you should consider using the residual strengths. The peak strength, which occurs at relatively small strains, is unconservative for analyses of slope stability. If the soil is structured or cemented (see Chapter 15) use of the critical state strength is likely to be overconservative: analyses of the stability of slopes in cemented soils are difficult and are beyond the scope of this book.

For slope stability analyses the factor of safety F_s should be applied to the soil strength so that the shear stresses mobilized in the soil, τ_a or τ'_a, are given by

$$\tau_a = \frac{s_u}{F_s} = s_{ua} \tag{20.2}$$

$$\tau'_a = \sigma' \frac{\tan \phi'_c}{F_s} = \sigma' \tan \phi'_a \tag{20.3}$$

where s_{ua} and ϕ'_a are allowable strength parameters. The idea here is to reduce the soil strength and then see that the slope is in equilibrium with the lower strengths.

The factor of safety should take account of uncertainties in the determinations of the loads (including the unit weight γ), the soil strengths and particularly the pore pressure or drainage conditions. The critical state strength of soils can be determined in laboratory tests on reconstituted samples with little uncertainty, although in practice the value may vary with depth through layered soils. The greatest uncertainty is in determination of steady state pore pressures in drained analyses or in the assumption of constant volume (and hence constant strength) in undrained analyses. There is no single value for F_s that can be recommended for slope stability calculations. Instead, you should investigate the consequences of changing the values of the loads, strengths and pore pressures. If you take the worst credible values for these parameters, values of F_s only slightly greater than unity will be adequate.

20.6 STABILITY OF INFINITE SLOPES

From now on I will examine the limiting stability of slopes with the critical state strengths s_u or ϕ'_c; to apply a factor of safety you can do the same calculations using s_{ua} or ϕ'_a obtained from Eqs (20.2) and (20.3). For slope stability calculations you can use the upper and lower bound method described in Chapter 18 or the limit equilibrium method described in Chapter 19. A simple but very useful case is for shallow sliding on a slip surface parallel to the slope, as illustrated in Fig. 20.3(b). The depth to the slip surface will be controlled by geological or groundwater conditions; a common case is where there is a mantle of soil over rock in a hillside and the slip surface is close to the interface between the soil and the rock.

(a) Undrained Loading

Figure 20.9(a) shows an infinite slope where the angle is an upper bound i_u with a mechanism of plastic collapse consisting of a slip surface through the soil at the rock level; there is a block of soil length l measured down the slope. The corresponding displacement diagram for an increment of displacement δw is shown in Fig. 20.9(b). For an infinitely long slope, the forces on any such block are the same as those on any other similar block and so the forces F_1 and F_2 are equal and opposite. From the geometry of Fig. 20.9(a) the weight of the block (for unit thickness normal to the page) is

$$W = \gamma H l \cos i_u \tag{20.4}$$

and from Fig. 20.9(b) the vertical component of displacement is

$$\delta v = \delta w \sin i_u \tag{20.5}$$

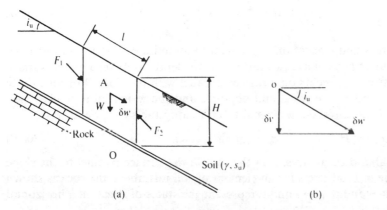

Figure 20.9 Mechanism of plastic collapse for an infinitely long slope for undrained loading.

(a)

(b)

Figure 20.10 Equilibrium state of stress for an infinitely long slope for undrained loading.

Hence, noting that the increments of work done by the equal and opposite forces F_1 and F_2 sum to zero, we have

$$\delta W = s_u l \, \delta w \tag{20.6}$$

$$\delta E = \gamma H l \cos i_u \, \delta w \sin i_u \tag{20.7}$$

and, equating $\delta W = \delta E$, an upper bound for the critical slope angle is given by

$$\sin i_u \cos i_u = \frac{s_u}{\gamma H} \tag{20.8}$$

and

$$i_u = \tfrac{1}{2} \sin^{-1} \frac{2 s_u}{\gamma H} \tag{20.9}$$

Figure 20.10(a) shows forces and stresses on an element in an infinite slope where the angle is a lower bound i_1. The state of stress increases linearly with depth from zero at the surface and the maximum shear stress $\tau = s_u$ occurs on a surface parallel with the slope. For an infinite slope, as before, the forces F_1 and F_2 are equal and opposite and the weight of a block of soil of length l is $W = \gamma H l \cos i_1$. Hence, resolving normal to and along the slope, we have

$$\sigma_s = \gamma H \cos^2 i_1 \qquad \tau_s = \gamma H \sin i_1 \cos i_1 \tag{20.10}$$

where σ_s and τ_s are the normal and shear stresses in the soil on the surface parallel to the slope at a depth H. The Mohr circle of total stress for an element of soil just above the rock is shown in Fig. 20.10(b). The pole is at P and points a and b represent the states of stress on a horizontal plane and on a plane parallel with the slope respectively; the angle subtended at the centre of the circle is $2i_1$. The Mohr circle just touches the undrained failure envelope and so the state of stress in the slope does not exceed the undrained failure criterion. From the geometry of Fig. 20.10(b), making use of Eq. (20.10), a lower bound for the critical slope angle is given by

$$\tan i_1 = \frac{\tau_s}{\sigma_s} = \frac{s_u}{\gamma H \cos^2 i_1} \tag{20.11}$$

and hence

$$i_1 = \tfrac{1}{2} \sin^{-1} \frac{2s_u}{\gamma H} \tag{20.12}$$

Comparing Eqs (20.9) and (20.12), the upper bound solution exactly equals the lower bound solution and so both must equal the exact solution. Hence the critical slope angle i_c for undrained loading of an infinite slope is given by

$$i_c = \frac{1}{2} \sin^{-1} \frac{2s_u}{\gamma H} \tag{20.13}$$

(b) Drained Loading—No Seepage

Figure 20.11(a) shows a mechanism of plastic collapse for an infinitely long slope whose angle to the horizontal is an upper bound i_u. The mechanism is a single slip surface at a depth z and there is a block of soil length l measured down the slope; as before, the forces F_1 and F_2 that act on the vertical sides are equal and opposite. The displacement diagram for an increment of displacement δw is shown in Fig. 20.11(b), where the direction of the increment of displacement makes an angle $\psi = \phi'_c$ to the slip surface.

For drained loading the increment of work done by the internal stresses for an increment of plastic collapse is $\delta W = 0$ and, noting that $F_1 = F_2$, the increment of work done by the external loads for dry soil is

$$\delta E = \delta v \, \gamma V \tag{20.14}$$

where $V = zl \cos i_u$ is the volume of the block. Hence, equating $\delta E = \delta W$, an upper bound is given by

$$\delta v \, \gamma V = 0 \tag{20.15}$$

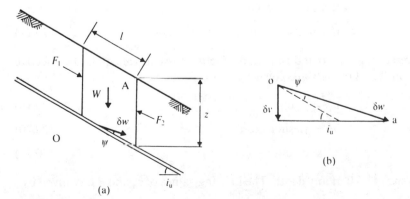

Figure 20.11 Mechanism of plastic collapse for an infinitely long slope in dry soil.

(a)

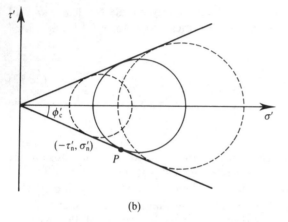

(b)

Figure 20.12 Equilibrium states of stress for an infinitely long slope in dry soil.

Since the volume V is non-zero, the upper bound is given by $\delta v = 0$ and hence, from the geometry of Fig. 20.11(b), an upper bound for the critical slope angle is given by

$$i_u = \phi'_c \tag{20.16}$$

Figure 20.12 shows an infinite slope whose angle with the horizontal is a lower bound i_l and a block of soil of length l measured down the slope and depth z measured vertically; the forces on the faces of the block are shown and, as before, the forces F_1 and F_2 are equal and opposite. Resolving normal to and parallel with the base AB the normal and shear forces N and T are

$$N = W \cos i_l = \gamma z l \cos^2 i_l \tag{20.17}$$

$$T = W \sin i_l = \gamma z l \sin i_l \cos i_l \tag{20.18}$$

For dry soil, where pore pressures are zero and total and effective stresses are equal, the effective normal and shear stresses on the plane AB are given by

$$\sigma'_n = \gamma z \cos^2 i_l \tag{20.19}$$

$$\tau'_n = \gamma z \sin i_l \cos i_l \tag{20.20}$$

and hence

$$\tau'_n = \sigma'_n \tan i_l \tag{20.21}$$

which is valid for all planes such as AB at any depth. The limiting values of τ'_n and σ'_n are given by

$$\tau'_n = \sigma'_n \tan \phi'_c \tag{20.22}$$

and hence a lower bound for the limiting slope angle is given by

$$i_1 = \phi'_c \qquad (20.23)$$

The Mohr circle of effective stress for the state of stress in an element on AB is shown in Fig. 20.12(b); the circles shown with broken lines correspond to the states of stress in elements above and below AB. All the Mohr circles just touch the drained failure envelope. The pole of the Mohr circle is at P and hence we may calculate the stresses on any other plane in the slope; in particular, the normal and shear stresses on vertical planes are equal in magnitude to those on planes parallel to the slope.

From Eqs (20.16) and (20.23) the upper and lower bounds are equal and hence the critical slope angle for dry soil is

$$i_c = \phi'_c \qquad (20.24)$$

(c) Drained Loading—Steady State Seepage

Figure 20.7(a) shows the flownet for steady state seepage parallel to the slope where the phreatic surface is a little below ground level. As water flows downhill there are additional seepage stresses that make the slope less stable and so you would expect to find that the critical slope angle i_c is less than ϕ'_c. Solutions can be found using the upper and lower bound methods but I shall show a limit equilibrium solution.

Figure 20.13(a) shows a mechanism consisting of a slip surface parallel to the slope at a depth z and the forces acting on a block length l down the slope and Fig. 20.13(b) is the polygon of forces. The forces on the slip surface are $T' = \tau'l$ and $N = \sigma l$, which is made up of $N' = \sigma'L$ and $U = ul$, where u is the pore pressure. From the force polygon,

$$T' = N \tan i_c = (N - U) \tan \phi'_c \qquad (20.25)$$

and the critical slope angle i_c is given by

$$\tan i_c = \tan \phi'_c \left(1 - \frac{U}{N}\right) \qquad (20.26)$$

The flownet for steady state seepage parallel to the slope is shown in Fig. 20.7(b) and, from Eq. (20.1), the pore pressure at a depth z is $u = \gamma_w mz \cos^2 i$. We have already calculated $\sigma_n = \gamma z \cos^2 i$ and hence, from Eq. (20.26), we have

$$\tan i_i = \left(1 - \frac{m\gamma_w}{\gamma}\right) \tan \phi'_c \qquad (20.27)$$

(a)

(b)

Figure 20.13 Limit equilibrium solution for an infinitely long slope with steady state seepage parallel with the slope.

For dry soil $m = 0$ and Eq. (20.27) reduces to Eq. (20.24). If the phreatic surface is at ground level $m = 1$ and noting that $\gamma \approx \frac{1}{2}\gamma_w$, Eq. (20.27) becomes

$$\tan i_c \approx \tfrac{1}{2} \tan \phi'_c \tag{20.28}$$

The solutions for the stability of infinite slopes given by Eqs (20.13), (20.23) and (20.28) are relatively simple. Notice that for the undrained slope the critical angle i_c is governed by the depth H of the slip surface; if this depth is relatively large the mechanism cannot be approximated to sliding parallel to the surface and the solution is no longer valid. For the drained case the critical angles for dry and submerged slopes are the same, $i_c = \phi'_c$ (because neither the unit weight nor the pore pressure appear in the final solution), but if there is steady state seepage parallel to the slope the critical slope angle is reduced. These results demonstrate, firstly, that water does not lubricate soil and, secondly, the very significant influence of pore pressures on slope stability.

20.7 STABILITY OF VERTICAL CUTS

A simple experiment with dry sand or sugar demonstrates that you cannot make a vertical cut in a drained soil. We can, however, make vertical cuts in soils that are undrained where the negative pore pressures generate positive effective stresses.

A simple collapse mechanism consisting of a single straight slip surface at an angle of 45° to the vertical is shown in Fig. 20.14(a) and Fig. 20.14(b) is the corresponding displacement diagram for an increment of displacement δw down the slip surface. From the geometry of Fig. 20.14(a), the length L of the slip surface and the volume V of the wedge (for unit thickness normal to the page) are given by

$$L = \sqrt{2}H_u \qquad V = \tfrac{1}{2}H_u^2 \tag{20.29}$$

where H_u is an upper bound for the height of the slope at collapse. From the geometry of Fig. 20.14(b) we have

$$\delta v = \frac{1}{\sqrt{2}}\, \delta w \tag{20.30}$$

The only external forces are those due to the self-weight of the sliding soil and

$$\delta W = s_u \sqrt{2} H_u\, \delta w \tag{20.31}$$

$$\delta E = \frac{1}{\sqrt{2}}\, \delta w\, \gamma \tfrac{1}{2} H_u^2 \tag{20.32}$$

(a) (b)

Figure 20.14 Mechanism of plastic collapse for a vertical cut slope for undrained loading.

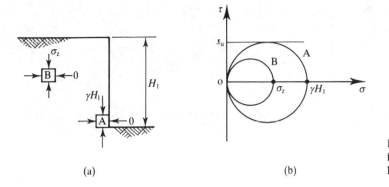

(a)   (b)

Figure 20.15 Equilibrium state of stress for a vertical cut slope for undrained loading.

Hence, equating $\delta W = \delta E$, an upper bound for the height of the cut slope at collapse is given by

$$H_u = \frac{4s_u}{\gamma} \tag{20.33}$$

For a lower bound Fig. 20.15(a) shows a state of stress in which shear stresses on vertical and horizontal planes are zero. The vertical and horizontal stresses are $\sigma_z = \gamma z$ and $\sigma_h = 0$, and these are principal stresses. Mohr circles of stress for the elements A and B in Fig. 20.15(a) are shown in Fig. 20.15(b). The Mohr circle A does not cross the undrained failure envelope when

$$\gamma H_1 = 2s_u \tag{20.34}$$

and hence a lower bound for the height of the cut is given by

$$H_1 = \frac{2s_u}{\gamma} \tag{20.35}$$

These upper and lower bound solutions are not really very close to one another and it is very difficult to obtain better solutions. The best solution, and the one that is commonly used in design, is

$$H_c = \frac{3.8s_u}{\gamma} \tag{20.36}$$

which is close to the upper bound given by Eq. (20.33). If the excavation is filled with water the critical height is given by

$$H_c = \frac{3.8s_u}{\gamma - \gamma_w} \tag{20.37}$$

Comparing Eqs (20.36) and (20.37), the critical height of a dry excavation is only about one-half that of an excavation filled with water.

You have probably noticed that the ground surface is often cracked and fissured, particularly near the top of a slope or excavation. Each vertical crack is like a small trench, as in Fig. 20.16, and the maximum depth of the crack is given by Eq. (20.36) or (20.37), depending on whether it is empty or filled with water. Notice that as pore pressures rise, the soil softens and weakens and the depth of the crack decreases; in the end, when the pore pressures are hydrostatic with a phreatic surface at ground level the cracks will have closed.

Figure 20.16 Stability of vertical cut slopes and vertical cracks filled with water.

(a)   (b)

20.8 ROUTINE SLOPE STABILITY ANALYSES

The most common procedure for slope stability analysis is to use the limit equilibrium method with a slip circle or a general curved slip surface. These methods were described in Chapter 19. For undrained loading (Sec. 19.4) the problem is statically determinate and the solution is relatively simple. For drained loading (Sec. 19.5) the problem is statically indeterminate and solutions using the method of slices require assumptions; there are a number of different solutions (e.g. Bishop, Janbu, Morgenstern and Price), each developed from different assumptions. In these solutions the calculations are largely repetitive and a number of standard computer programs are available for slope stability analysis.

For slopes with relatively simple geometries, standard solutions are available in the form of non-dimensional tables and charts. These are very useful for preliminary design studies.

(a) Stability Numbers for Undrained Loading

The solution for an infinite slope for undrained loading was given by Eq. (20.13), which can be rewritten as

$$H_c = \frac{2}{\sin 2i} \frac{s_u}{\gamma} \tag{20.38}$$

or

$$H_c = N_s \frac{s_u}{\gamma} \tag{20.39}$$

where N_s is a stability number that depends principally on the geometry of the slope.

Figure 20.17(b) shows a more general case where strong rock occurs at a depth $n_d H$ below the top ground level and Fig. 20.17(a) shows values of the stability number N_s in terms of the slope angle i and the depth factor n_d. The data in Fig. 20.17 are taken from those given by Taylor (1948, p. 459) and were obtained from the limit equilibrium slip circle method.

(b) Stability Numbers for Drained Loading

The safe slope angle for drained loading with steady state seepage is obtained from Eq. (20.27) substituting the allowable friction angle ϕ'_a for ϕ'_c and is given by

$$\tan i = \tan \phi'_a \left(1 - \frac{U}{N} \right) \tag{20.40}$$

where $U = \gamma_w z \cos^2 i$ and $N = \gamma_z \cos^2 i$. From Eqs (20.3) and (20.40), noting that $\sigma_z = \gamma z$, we have

$$F_s = \frac{\tan \phi'_c}{\tan i} \left(1 - \frac{u}{\sigma_z} \sec^2 i \right) \tag{20.41}$$

Figure 20.17 Stability numbers for undrained loading. (After Taylor, 1948.)

Figure 20.18 Stability numbers for drained loading. (After Bishop and Morgenstern, 1960.)

Equation (20.41) can be written as

$$F_s = m - nr_u \tag{20.42}$$

where m and n are stability numbers that depend on the geometry of the slope and on the friction angle ϕ' and $r_u = u/\sigma_z$ is a pore pressure coefficient. Figure 20.18 shows values for the stability numbers m and n for simple slopes calculated by Bishop and Morgenstern (1960) from slip circle analysis using the method of slices. For a particular slope an average value of r_u must be estimated from a steady state seepage flownet and from the position of the critical slip circle: in many practical cases r_u is taken as about $\frac{1}{3}$.

20.9 BEHAVIOUR OF SIMPLE EXCAVATIONS

All the features of slope stability described in the previous section can be observed by digging a hole in the beach. What you will see is illustrated in Fig. 20.19. In the dry sand at the surface

Figure 20.19 Stability of a simple excavation.

the slope angle is ϕ'_c. In the unsaturated sand above the water table the pore pressures are negative and it is possible to excavate a vertical cut BC. The cut will fail if the depth exceeds the critical height H_c; this is given by Eq. (20.13) where s_u can be found from an unconfined compression test carried out on a sand-castle at the same density and water content (see Sec. 9.6). The vertical cut cannot be continued below the water table C where the pore pressures are zero. (The cut often fails just above the water table where the sand is saturated and the negative pore pressures are small.) Notice that pore pressures behind the cut BC are negative so the face should look dry.

You know that it is very difficult to dig the hole below the water table. If you excavate slowly there will be steady state seepage so the angle of the slope CD will be about $\frac{1}{2}\phi'_c$. If you can excavate below water so there is no seepage the angle of the slope DE will be about ϕ'_c. In practice seepage into the excavation along CD usually causes erosion due to piping (Sec. 17.6) and you cannot dig much below the water table.

When you do this experiment remember that the factor of safety of the vertical cut BC is probably reducing with time and you must be very careful that it does not collapse on you. You should also observe what happens to your hole as the tide comes in or as the sun shines on to the face BC.

20.10 SUMMARY

1. Slopes fail as soil moves on slip surfaces and there are several possible mechanisms depending on the ground and groundwater conditions.
2. Immediately after excavation or filling pore pressures are reduced and, as time passes, pore pressures rise, effective stresses reduce and the safety of a slope deteriorates.
3. For slope stability calculations the factor of safety accounts for uncertainties in the determination of the soil parameters and the analyses. For routine analyses the critical state strength will give safe designs with factors of safety accounting for uncertainties in the pore pressures. If previous landsliding has occurred the strength may have reduced to the residual before construction starts.
4. Slope stability calculations can be done using the upper and lower bound methods or the

limit equilibrium method; preliminary designs can be carried out making use of routine stability numbers.

WORKED EXAMPLES

Example 20.1: Undrained slope stability Figure 20.20 shows the geometry of a simple slope. From Eq. (20.39) and replacing s_u with s_u/F_s,

$$F_s = \frac{N_s s_u}{\gamma H}$$

From Fig. 20.17, for $i = 20°$ and $n_d = \infty$ we have $N_s = 5.5$ and

$$F_s = \frac{5.5 \times 40}{20 \times 5} = 2.2$$

Notice that this is rather less than the result $F_s = 2.98$ obtained for Example 19.3, indicating that the slip circle in Fig. 19.18 was not the critical one.

$s_u = 40\,\text{kPa}$
$\phi' = 30°$
$\gamma = 20\,\text{kN/m}^3$

$20°$ 5m

Figure 20.20

Example 20.2: Drained slope stability For drained loading of the slope in Fig. 20.20, from Eq. (20.42),

$$F_s = m - nr_u$$

For $i = 20°$ the gradient is 2.75:1 and, from Fig. 20.18, for $\phi' = 30°$ we have $m \approx 1.6$ and $n \approx 1.8$. Taking a characteristic value for $r_u = 0.3$,

$$F_s = 1.6 - (0.3 \times 1.8) = 1.06$$

Near the toe of the slope the flowlines will be approximately parallel to the slope and the phreatic surface is close to ground level. From Eq. (20.27), replacing $\tan \phi'_c$ with $\tan \phi'_c/F_s$,

$$F_s = \left(1 - \frac{m\gamma_w}{\gamma}\right)\frac{\tan \phi'_c}{\tan i}$$

If the phreatic surface is at ground level $m = 1$,

$$F_s = \left(1 - \frac{10}{20}\right)\frac{\tan 30°}{\tan 20°} = 0.80$$

and local instability will occur near the toe. In order to stabilize the slope the drain in Fig. 19.19 should be lowered to reduce the value of m.

REFERENCES

Bishop, A. W. and N. R. Morgenstern (1960) 'Stability coefficients for earth slopes', *Geotechnique*, **10**, 129–150.
Taylor, D. W. (1948) *Fundamentals of Soil Mechanics*, Wiley, New York.

FURTHER READING

Atkinson, J. H. (1981) *Foundations and Slopes*, McGraw-Hill, London.
Bromhead, E. N. (1986) *The Stability of Slopes*, Surrey University Press, London.

CHAPTER TWENTY-ONE

EARTH PRESSURES AND STABILITY OF RETAINING WALLS

21.1 INTRODUCTION

Retaining walls are used to support slopes and vertical cuts that are too steep or too deep to remain stable if unsupported. The principal characteristics of a retaining structure are illustrated in Fig. 21.1. The wall is a structural member that acts as a beam with various loads on either side. Slender walls are embedded into the ground below the excavation level and they may be supported by props or anchors. Thick heavy gravity walls derive their resistance principally from the shear stresses between the soil and the base of the wall. During excavation (or filling on the high side) slender walls will tend to move and bend as indicated as the earth pressures develop. Walls move towards the passive side and away from the active side.

The development of earth pressure with displacement is illustrated in Fig. 21.2. In Fig. 21.2(a) a wall supported by a force P retains soil where the horizontal total stress is σ_h; obviously the stresses and the force must be in equilibrium. If P is increased the wall moves towards the passive side with displacements δ_p and the horizontal stresses increase, as shown in Fig. 21.2(b); if P is decreased the wall moves towards the active side with displacements δ_a and the horizontal stresses decrease. If the movements are sufficient the horizontal stresses reach the limiting values of the passive pressure σ_p and the active pressure σ_a. If there is no movement the horizontal stress σ_0 is the earth pressure at rest, corresponding to K_0 (see Sec. 8.5).

The design of retaining walls requires calculation of the active and passive earth pressures, the depth of embedment to ensure overall stability and the loads in the wall and in any props or anchors. Other things to be considered are ground movements and groundwater seepage.

Figure 21.1 Characteristics of a retaining wall.

275

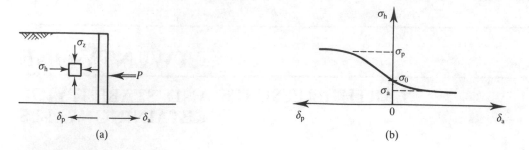

Figure 21.2 Development of active and passive pressures with displacement.

21.2 TYPES OF RETAINING STRUCTURE

There are a number of different types of retaining wall and the principal ones are illustrated in Fig. 21.3. Figure 21.3(a) shows a simple cantilever wall where all the support comes from the passive earth pressures. Figure 21.3(b) and (c) illustrates simple propped and anchored walls respectively. Figure 21.3(d) shows a gravity wall where the resistance comes from shear stresses between the ground and the base of the wall. In Fig. 21.3(e) the wall supports the sides of an excavation and in Fig. 21.3(f) the wall supports fill.

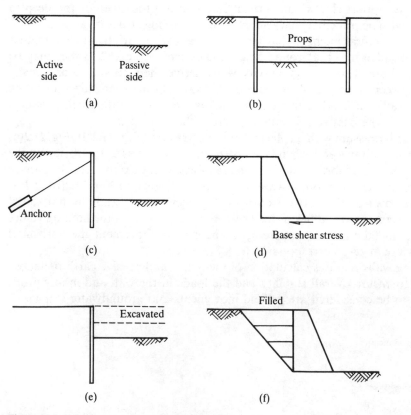

Figure 21.3 Principal types of retaining structure.

Permanent walls are used to support highway cuttings, bridge abutments, basements, dock and harbour walls and so on, while temporary retaining walls are used extensively during construction to support excavations and to provide dry working conditions in coffer dams. Gravity walls are usually of masonry or mass concrete but could also be made from gabions (wire baskets about 0.5 to 1 m cube filled with soil or rock). Slender walls are steel or reinforced concrete. Steel sheet piles are usually driven into the ground while slender concrete walls are usually cast *in situ* as rectangular diaphragm panels or as interlocking or touching cylindrical piles.

21.3 FAILURE OF RETAINING WALLS

Retaining walls can fail in a number of different ways. Figure 21.4 illustrates typical failure in the soil where the wall itself remains intact and Fig. 21.5 illustrates typical failures of the structural elements. The walls in Fig. 21.4(a) and (b) are failing because there are very large distortions in the soil in front of and behind the wall. In Fig. 21.4(c) and (d) a gravity wall may fail by sliding, overturning or by exceeding the limiting bearing pressure at the toe. In Fig. 21.4(e) any retaining wall may fail by slipping below the wall but this is really a problem in slope stability (see Chapter 20). In Fig. 21.4(f) the base of an excavation may fail by piping and erosion

Figure 21.4 Mechanisms of failure of retaining walls.

Figure 21.5 Structural failures of retaining walls.

due to seepage or by movement of the soil. Figure 21.5 illustrates structural failures of the wall or an anchor or buckling of props.

21.4 STRESS CHANGES IN SOIL NEAR RETAINING WALLS

It is helpful to consider the total and effective stress paths in soil near retaining walls during and after construction to examine whether the undrained or long-term drained cases are most critical. For retaining walls it is necessary to separate those loaded by excavation from those loaded by filling. (Note that I am continuing to use loading to mean an increase of shear stress irrespective of what happens to the normal stresses.)

Figure 21.6(a) shows a retaining wall loaded by excavation. For both the elements shown on the critical slip surfaces, one on the active side and one on the passive side, the shear stresses increase while the mean normal total stresses decrease. The total and effective stress paths are shown in Fig. 21.6(b); these are like those for a slope, shown in Fig. 20.4. The effective stress path $A' \rightarrow B'$ corresponds to undrained loading: the exact path will depend on the characteristics of the soil and on its initial overconsolidation ratio, as discussed in Chapter 11.

As shown in Fig. 21.6(b), the pore pressure immediately after construction u_i is less than the final steady state pore pressure u_c and so there is an initial excess pore pressure which is negative. As time passes the total stresses remain approximately unchanged at B (they will change a little as the total stresses redistribute during consolidation, although there is no more excavation) but the pore pressures rise. The effective stress path is $B' \rightarrow C'$, which corresponds to swelling and a reduction in the mean normal effective stress. The final state at C' corresponds to a steady state pore pressure after swelling u_c.

The wall will fail in some way if the states of all elements along the slip surfaces in Fig. 21.6(a) reach the critical state line; if B' reaches the critical state line the wall fails during undrained excavation and if C' reaches the line the wall fails some time after construction. The

(a)

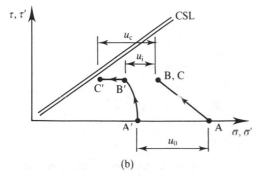

(b)

Figure 21.6 Changes of stress and pore pressure for a wall retaining an excavation.

distance of the effective stress point B′ or C′ from the critical state line is a measure of the factor of safety against collapse and Fig. 21.6(b) demonstrates that the factor of safety of a retaining wall supporting an excavation will decrease with time. This is the same as for a slope, discussed in Sec. 20.3. We could also trace the state paths for failing walls as we did for failing slopes, but this is not really relevant as retaining walls should not be allowed to fail.

Figure 21.7(a) shows a wall embedded in soil and retaining coarse-grained fill. In this case the shear and normal stresses on typical elements on a slip surface both increase. Total and effective stress paths are shown in Fig. 21.7(b). The effective stress path for undrained loading is A′ → B′ and this is the same as that in Fig. 21.6(b), but the total stress path A → B and the initial pore pressures are different. In particular, the initial pore pressure u_i is greater than the final steady state pore pressure, so the initial excess pore pressure is positive. As time passes the pore pressures decrease as the soil consolidates and the effective stress path is B′ → C′. The effective stress point is moving away from the critical state line so the factor of safety of a wall retaining fill increases with time.

The analyses and the stress paths shown in Figs 21.6 and 21.7 are simplified and idealized and ignore a number of important aspects such as the installation of the wall into the ground. They do, however, illustrate the general features of the behaviour of retaining walls during and after construction. Notice particularly the fundamental difference between the long-term behaviour of walls supporting excavations and walls retaining fill: the one becomes less safe with time as the soil softens and weakens and the other becomes safer with time as the soil consolidates and strengthens.

21.5 INFLUENCE OF WATER ON RETAINING WALLS

Water influences the loading on retaining walls in a number of fundamentally different ways; the most important of these are illustrated in Fig. 21.8. Figure 21.8(a) shows a coffer dam wall

(a)

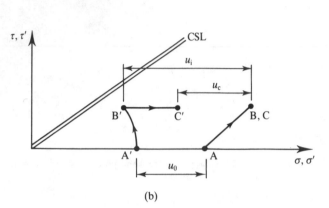

(b)

Figure 21.7 Changes of stress and pore pressure for a wall retaining granular fill.

embedded in soil and retaining water. The free water applies a total stress P_w to the wall where

$$P_w = \tfrac{1}{2}\gamma_w H_w^2 \tag{21.1}$$

Figure 21.8(b) shows a wall retaining soil. There is water in the excavation which applies a total stress P_w and the wall is supported by a single prop with a load P_a. (It is assumed that the prop is placed so that the wall does not rotate.) The total stress applied to the soil arises from the sum of P_w and P_a: notice that this is the same whether the soil is drained or undrained and whether the wall is impermeable or leaky.

Figure 21.8(c) shows a wall supporting a coarse-grained soil which is loaded drained. The toe of the wall is embedded in relatively impermeable clay and the excavation is dry. If the wall is impermeable it acts as a dam and the pore pressures are everywhere hydrostatic. The pore pressures apply a force P_w to the wall in addition to the horizontal effective stresses. The strength of the soil on the slip surface shown is reduced by the influence of the pore pressures lowering the effective stresses. Figure 21.8(d) shows the same wall but with a drain near the toe and a sketched flownet for steady state seepage. It is obvious that the force P_a required to support the wall has been significantly reduced: there are no water pressures acting directly on the wall and the effective stresses, and the strength, on the slip surfaces are greater because the pore pressures are less. The example illustrates the importance of providing adequate drainage for retaining walls.

Figure 21.8(e) shows steady state seepage into a pumped coffer dam. (The flownet is similar to the one shown in Fig. 17.7.) At the bottom of the coffer dam, along AB, there is upward seepage and the possibility of instability due to piping and erosion, discussed in Sec. 17.6. Piping will occur when the hydraulic gradient $i = \delta P/\delta s$ becomes close to unity. For the example illustrated, δP over the last element of the flownet is $\Delta P/7$ (because there are seven equipotential drops in the flownet) and the size of the last element δs can be determined by measurement from a scaled diagram.

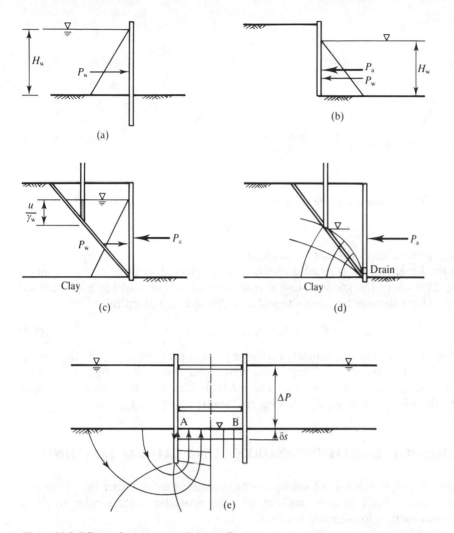

Figure 21.8 Effects of water on retaining walls.

21.6 CALCULATION OF EARTH PRESSURES—DRAINED LOADING

As a retaining wall moves the horizontal stresses change, as illustrated in Fig. 21.2, and when they reach the limiting active or passive pressures the soil has reached its critical state. The active and passive pressures can be calculated using upper and lower bound and limit equilibrium methods and, as always, it is necessary to distinguish between drained and undrained loading.

A limit equilibrium solution for the active force on a wall retaining dry soil was found in Sec. 19.3. The mechanism and the polygon of forces were shown in Fig. 19.6 and the solution is

$$P_a = \tfrac{1}{2}\gamma H^2 \tan^2(45° - \tfrac{1}{2}\phi'_c) \tag{21.2}$$

Assuming that the effective active pressure σ'_a increases linearly with depth the earth pressures corresponding to this limit equilibrium solution are

$$\sigma'_a = \sigma'_z \tan^2(45° - \tfrac{1}{2}\phi'_c) = K_a \sigma'_v \tag{21.3}$$

where σ'_z is the vertical effective stress and K_a is called the active earth pressure coefficient. It is

Figure 21.9 Earth pressures on a rough wall with a sloping face and with sloping ground.

quite easy to show that the solution for the passive pressure is

$$\sigma'_p = \sigma'_z \tan^2(45° + \tfrac{1}{2}\phi'_c) = K_p\sigma'_v \tag{21.4}$$

where K_p is called the passive earth pressure coefficient.

These solutions are for a smooth vertical wall with a level ground surface. A more general case is shown in Fig. 21.9 where the ground surface and the back of the wall are both inclined and the wall is rough. Shear stresses between the soil and the wall are given by

$$\tau'_s = \sigma'_n \tan \delta'_c \tag{21.5}$$

where σ'_n is the normal stress for the appropriate active or passive pressure and δ'_c is the critical angle of wall friction. Obviously $0 < \delta'_c < \phi'_c$ and a value commonly taken for design is $\delta'_c = \tfrac{2}{3}\phi'_c$. The general case was considered in Sec. 19.3 (see Fig. 19.7) in Chapter 19. Tables and charts are available giving values for K_a and K_p for various combinations of ϕ'_c, δ'_c, α and β.

21.7 CALCULATION OF EARTH PRESSURES—UNDRAINED LOADING

Active and passive pressures for undrained loading can be calculated using either the upper and lower bound methods or the limit equilibrium method. The procedures are similar to those described in the previous section for drained loading.

A limit equilibrium solution for the active pressures on a smooth wall was obtained in Sec. 19.3 from the limit equilibrium method using the Coulomb wedge analysis (see Fig. 19.4). The solution was

$$P_a = \tfrac{1}{2}\gamma H^2 - 2s_u H \tag{21.6}$$

and, assuming that the stresses increase linearly with depth,

$$\sigma_a = \gamma z - 2s_u \tag{21.7}$$

It is relatively simple to show that the passive pressure for undrained loading is given by

$$\sigma_p = \gamma z + 2s_u \tag{21.8}$$

These expressions for active and passive earth pressures for undrained loading can be written as

$$\sigma_a = \sigma_z - K_{au}s_u \tag{21.9}$$

$$\sigma_p = \sigma_z + K_{pu}s_u \tag{21.10}$$

where K_{au} and K_{pu} are earth pressure coefficients for undrained loading.

The solutions with $K_{au} = K_{pu} = 2$ are for a smooth vertical wall with a level ground surface.

(a) With a tension crack

(b) With a surface surcharge

Figure 21.10 Active forces on walls—undrained loading.

Tables and charts are available giving values for K_{au} and K_{pu} for other cases including rough walls where the shear stress between the soil and the wall is s_w.

From Eq. (21.7) the active earth pressure for undrained loading appears to become negative (i.e. in tension) when

$$z < \frac{2s_u}{\gamma} \tag{21.11}$$

This is impossible as the soil is not glued to the wall and a tension crack opens up as shown in Fig. 21.10(a). This is the same kind of tension crack as found near the top of slopes (see Sec. 20.7) and the critical depth H_c of a water-filled crack is

$$H_c = \frac{2s_u}{\gamma - \gamma_w} \tag{21.12}$$

If the crack is not filled with water put $\gamma_w = 0$ into Eq. (21.12). Notice that the position of the active force P_a has been lowered and if the crack is filled with water it is free water (not pore water) and applies a total stress to the wall. If there is a surface stress q as shown in Fig. 21.10(b), the tension crack will close entirely when $q = 2s_u$.

Compare Eqs (21.9) and (21.10) for undrained loading with Eqs (21.3) and (21.4) for drained loading. For undrained loading the earth pressure coefficients are expressed as a difference $(\sigma_h - \sigma_z)$ while for drained loading they are a ratio (σ'_h/σ'_z). This is a consequence of the fundamental difference between the basic equations for drained and undrained strength.

21.8 OVERALL STABILITY

The forces on a retaining wall arise from the active and passive earth pressures, from free water pressures and from loads in props and anchors. For overall stability the forces and moments arising from these pressures must be in equilibrium. For the simplified example shown in Fig. 21.11,

$$P + \int_0^{H_w} \sigma_w \, dz = \int_0^H \sigma_h \, dz \tag{21.13}$$

where the integrals are simply the areas under the pressure distribution diagrams. In order to take moments it is necessary to determine the moment arm of each force; the line of action of a force is through the centre of area of each pressure distribution diagram.

The best way to avoid making mistakes is to set up a table and draw the distribution of

Figure 21.11 Influence of free water on the loads on a retaining wall.

earth pressure with depth, as shown in Table 21.1 in Example 21.1 below. This gives calculations for the horizontal stresses on a wall bedded into sand over clay as shown in Fig. 21.16(a). The calculations use Eqs (21.3) and (21.4) for the stresses in the sand and Eqs (21.9) and (21.10) for the stresses in the clay: in the free water the horizontal and vertical total stresses are equal. Notice how the pore pressures come into the calculations in the drained sand but not in the undrained clay. There is a step in the earth pressures at the sand–clay junction, so it is necessary to calculate separately the stresses just in the sand and just in the clay.

Overall, a wall is considered to be stable if the forces and moments are in equilibrium and this is examined by resolving horizontally and taking moments about a convenient point. In most analyses the variable (or unknown) is the depth of embedment, which is increased until a suitable margin of safety is achieved. Selection of factors of safety for a retaining wall design is very difficult and will be considered in a later section: for the present I will simply consider the overall stability of a retaining wall at the point of collapse, such that the horizontal stresses are everywhere the full active and passive pressures. It is necessary to consider propped or anchored walls, cantilever walls and gravity walls separately.

(a) Anchored or Propped Walls

Figure 21.12 shows a simple propped wall with depth of embedment d. The active and passive pressures are as shown and from these the magnitudes P and depths z of the active and passive forces are calculated as described in the previous section. Taking moments about P, the line of action of the prop forces, the wall is stable if

$$P_a z_a = P_p z_p \tag{21.14}$$

Resolving horizontally, the prop or anchor force P is given by

$$P = P_a - P_p \tag{21.15}$$

Notice that all the terms in Eq. (21.14) depend on the (unknown) depth of penetration d and

Figure 21.12 Forces on a propped wall.

Figure 21.13 Forces on cantilever walls.

solutions are most easily found by trial and error, adjusting d until Eq. (21.14) is satisfied. In Fig. 21.12 the toe of the wall rotates and translates and this is known as the free earth support condition. If the depth d is very large the toe of the wall will not translate or rotate; this is known as the fixed earth support condition.

(b) Cantilever Walls

If there is no prop or anchor it is impossible to satisfy moment and force equilibrium at the same time with only the two forces P_a and P_p. Stiff cantilever walls fail by rotation about a point some way above the toe, as shown in Fig. 21.13, and this system of forces can satisfy moment and force equilibrium. It is convenient to replace the forces below the point of rotation by a single force Q, as shown in Fig. 21.13(b). Taking moments about Q the wall is stable if

$$P_a h_a = P_p h_p \tag{21.16}$$

which gives the unknown depth of penetration d. In order to allow the wall below the point of rotation to mobilize the pressures shown in Fig. 21.13(a), the depth d is usually increased by 20 per cent.

The walls shown in Figs 21.12 and 21.13 can be considered as beams carrying concentrated and distributed loads. The shear forces and bending moments in the wall can be calculated using the standard analyses for beams.

(c) Gravity Walls

Gravity walls may fail by sliding, by overturning or by failure of the soil at the toe, as illustrated in Figs 21.4 and 21.14. Figure 21.14(a) shows a wall failing by sliding along its base and $P_a = T$. For undrained loading,

$$T = s_w B \tag{21.17}$$

where s_w is the undrained shear strength between the soil and the base of the concrete wall. For drained loading,

$$T = (W - U) \tan \delta'_c \tag{21.18}$$

where δ'_c is the angle of shearing resistance between the soil and the wall and U is the force

(a) Sliding on the base

(b) Overturning

(c) Bearing capacity failure

Figure 21.14 Equilibrium of gravity retaining walls.

due to pore pressures acting over the base area B. The wall cannot overturn provided that the normal stress at the upstream edge remains positive (i.e. in compression) and it can be shown by simple statics that this requires that the resultant R passes through the middle third of the base, as shown in Fig. 21.14(b). The resulting triangular distribution of normal stress shown in Fig. 21.14(c) implies that the maximum stress at the toe is given by

$$q = \frac{2W}{B} \tag{21.19}$$

The possibility of failure of the foundation due to excessive bearing pressure is really a problem of bearing capacity and is discussed in Chapter 22.

21.9 CHOICES OF SOIL STRENGTH AND FACTOR OF SAFETY

So far I have described analyses for overall stability of retaining walls based on the ultimate critical state strengths s_u or ϕ'_c with no factor of safety. These situations correspond to relatively large ground movements and, in practice, a factor of safety is applied.

There is no clearly defined method for applying a factor of safety in retaining wall design and factors are applied for different purposes:

1. To ensure an adequate margin of safety against failure in the soil (assuming that the wall itself and the anchors and props do not fail).
2. To limit prop and anchor loads and shear force and bending moments in the wall to permissible values.
3. To limit ground movements.

The issues are far too complicated for this book and you will have to consult other books for

Figure 21.15 Approximate distribution of earth pressures on a retaining wall at working loads.

details. If you are concerned only with overall stability it is convenient to apply a safety factor to the critical state soil strength, as described in Sec. 20.5 dealing with factors of safety for slopes. This method may not, however, lead to satisfactory solutions for ground movements or structural loads.

A major difficulty is that the distributions of active and passive earth pressures on safely designed walls are often very different to those on a wall close to ultimate failure, which are the basis of a design. Figure 21.15 shows the likely distribution of stress on a propped wall where the depth of the toe is greater than that required for limiting stability. The broken lines correspond to the zero movement earth pressure at rest K_0 condition (see Sec. 8.5) and the final stresses will tend to these lines at the bottom of the wall where the displacements are small. The chain dotted lines for K_a and K_p correspond to full active and passive pressures reached after significant ground movements. The full lines represent possible distributions of earth pressure; these will depend significantly on the ability of the wall to bend (i.e. on its flexibility). These different distributions of earth pressure are likely to have a major influence on the magnitudes of the shear forces and bending moments in a wall.

21.10 SUMMARY

1. Retaining walls are used to support slopes that are too high or too steep to remain stable if unsupported or to limit ground movements. There are a number of different kinds of retaining wall. They can fail in different ways including slipping in the soil, failure of the wall itself and failure of props or anchors.
2. As a wall moves away from the soil the horizontal stresses are active pressures and as it moves towards the soil they are passive pressures. For drained loading on smooth walls these are

$$\sigma'_a = \sigma'_z \tan^2(45° - \tfrac{1}{2}\phi'_c) = K_a \sigma'_v \tag{21.3}$$

$$\sigma'_p = \sigma'_z \tan^2(45° + \tfrac{1}{2}\phi'_c) = K_p \sigma'_v \tag{21.4}$$

where K_a is the active earth pressure coefficient and K_p is the passive earth pressure coefficient. For undrained loading on smooth wall active and passive pressures are

$$\sigma_a = \sigma_z - K_{au} s_u \tag{21.9}$$

$$\sigma_p = \sigma_z + K_{pu} s_u \tag{21.10}$$

where K_{au} and K_{pu} are earth pressure coefficients for undrained loading.

3. For walls retaining excavated slopes pore pressures rise with time and the safety deteriorates, but for walls retaining coarse grained fill the excess pore pressures developed in the foundations during construction will generally decrease with time.

4. The depth of the toe of a wall below the base of the excavation must be sufficient to ensure overall stability (with an appropriate margin of safety). Overall stability is examined by considering the statical equilibrium of the forces due to the active and passive earth pressures and the loads in props and anchors. Different calculations are required for cantilever and propped walls.

WORKED EXAMPLES

Example 21.1: Calculation of active and passive earth pressures Figure 21.16 shows a 10 m high wall retaining layers of sand and clay. The active and passive total stresses in the drained sand are

$$\sigma_a = \sigma'_a + u = \sigma'_z K_a + u = (\sigma_z - u)K_a + u$$

$$\sigma_p = \sigma'_p + u = \sigma'_z K_p + u = (\sigma_z - u)K_p + u$$

where $K_a = \tan^2(45° - \tfrac{1}{2}\phi'_c)$ and $K_p = \tan^2(45° + \tfrac{1}{2}\phi'_c)$ and, for $\phi' = 30°$, $K_p = 1/K_a = 3$. The total active and passive stresses in the undrained clay are

$$\sigma_a = \sigma_z - K_{au}s_u$$

$$\sigma_p = \sigma_z + K_{pu}s_u$$

where, for a smooth wall, $K_{au} = K_{pu} = 2$. The variations of σ_a and σ_p with depth are given in Table 21.1; to calculate active and passive pressures in layered soils and where there are pore pressures it is convenient to tabulate the calculations in this way. Notice that the stresses at the base of the sand are not the same as the stresses at the top of the clay. Figure 21.17 shows the variations of active and passive total pressures with depth.

Figure 21.16

Table 21.1

(a) Active side

Depth (m)	Soil	σ_z (kPa)	u (kPa)	σ'_z (kPa)	σ'_a (kPa)	σ_a (kPa)
0	Sand	80	0	80	27	27
2	Sand	120	0	120	40	40
7	Sand	220	50	170	57	107
7	Clay	220				140
10	Clay	280				200

(b) Passive side

Depth (m)	Soil	σ_z (kPa)	u (kPa)	σ'_z (kPa)	σ'_p (kPa)	σ_p (kPa)
2	Water	0	0	0	0	0
5	Water	30	30	0	0	30
5	Sand	30	30	0	0	30
7	Sand	70	50	20	60	110
7	Clay	70				150
10	Clay	130				210

σ_n (kPa)

Figure 21.17

Example 21.2: Depth of a propped wall   Figure 21.18(a) shows a wall propped at the top retaining dry sand. The unknown depth of penetration is d. For a factor of safety $F_s = 1.6$ the allowable angle of friction, given by $\tan \phi'_a = \tan \phi'_c / F_s$ is $\phi'_a = 20°$. Hence, from Eqs (21.3) and (21.4),

$$K_a = \tan^2(45° - \tfrac{1}{2}\phi'_a) = \tan^2 35° = 0.5$$

$$K_p = \tan^2(45° + \tfrac{1}{2}\phi'_a) = \tan^2 55° = 2.0$$

With the depth H measured from the ground level on either side of the wall and making use of Eq. (21.14),

$$P_a = \tfrac{1}{2}\gamma H^2 K_a = \tfrac{1}{2} \times 20 \times (5 + d)^2 \times \tfrac{1}{2} = 5(5 + d)^2 \quad \text{kN}$$

$$P_p = \tfrac{1}{2}\gamma H^2 K_p = \tfrac{1}{2} \times 20 \times d^2 \times 2 = 20d^2 \quad \text{kN}$$

The distributions of active and passive earth pressures and the active and passive forces are shown in Fig. 21.18(b). Taking moments about the top of the wall and noting that the forces P_a and P_p act at the centres of the triangular areas (i.e. $\tfrac{1}{3}H$ above the base),

$$5(5 + d)^2 \times \tfrac{2}{3}(5 + d) = 20d^2 \times (5 + \tfrac{2}{3}d)$$

and, solving by trial and error, or otherwise,

$$d = 4.0 \text{ m}$$

With this value of d we have $P_a = 405$ kN and $P_p = 320$ kN. Hence, resolving horizontally, from Eq. (21.15) the force in the prop is,

$$P = 405 - 320 = 85 \text{ kN}$$

(a)   (b)

(c)

Figure 21.18

Example 21.3: Depth of a cantilever wall   If the wall in Fig. 21.18(a) is not propped it acts as a cantilever and the forces on the wall are shown in Fig. 21.18(c). From Eq. (21.16), taking moments about the toe where the force Q acts,

$$5(5 + d)^2 \times \tfrac{1}{3}(5 + d) = 20d^2 \times \tfrac{1}{3}d$$

and, solving that by trial and error, or otherwise,

$$d = 8.5 \text{ m}$$

To provide sufficient length to mobilize the force Q, the wall depth should be increased by 20 per cent so the required depth of penetration is about 10 m.

FURTHER READING

Atkinson, J. H. (1981) *Foundations and Slopes*, McGraw-Hill, London.
Clayton, C. R. I. and J. Milititsky (1986) *Earth Pressure and Earth Retaining Structures*, Surrey University Press, London.
Heyman, J. (1972) *Coulomb's Memoir on Statics*, Cambridge University Press, Cambridge.
Kerisel, J. and E. Absi (1990) *Active and Passive Pressure Tables*, Balkema, Rotterdam.
Padfield, C. J. and R. J. Mair (1984) *Design of Retaining Walls Embedded in Stiff Clays*, CIRIA, Report 104, London.
Rankine, W. J. M. (1857) 'On the stability of loose earth', *Phil. Trans. R. Soc.*, **147**, 9–28.

CHAPTER TWENTY-TWO

BEARING CAPACITY AND SETTLEMENT OF SHALLOW FOUNDATIONS

22.1 TYPES OF FOUNDATIONS

Any structure that is not flying or floating rests on or in the ground and the base of the structure and the soil together make up the foundation. Buildings and embankments must have foundations and so must vehicles and people. The criteria for the design of a foundation are that the settlements should be limited so that the building does not become damaged, vehicles can still move about and you do not lose your boots in mud. All foundations settle because nothing (not even tarmac or rock) is absolutely rigid, but obviously some settle more than others; look at the Tower of Pisa for instance. When you walk across the beach and leave a footprint it is simply a mark of the settlement of a foundation and so too is a tyre track.

In civil engineering foundations are shallow, deep or piled, as illustrated in Fig. 22.1. (The distinction $D/B = 1$ to 3 for a deep foundation is made for convenience.) We know that, in general, the strength and stiffness of homogeneous soil increases with depth (because mean effective stresses increase with depth) and so one advantage of a deep foundation and a pile is

(a) Shallow

(b) Deep $D/B = 1$ to 3

(c) Piled

Figure 22.1 Types of foundation.

Figure 22.2 Loads and stresses on a foundation.

that they are founded in stronger and stiffer soil; often the tip of a pile rests on very stiff and strong soil or rock. Another advantage is that shear stresses between the soil and the sides of a deep foundation or a pile contribute to the load capacity; in a shallow foundation the contribution of the side shear stresses is negligible.

The characteristics of a typical foundation are illustrated in Fig. 22.2. The weight of the foundation is W and it supports a load F. The base width is B; for unit length out of the page this is the base area, so the bearing pressure q is

$$q = \frac{F + W}{A} \tag{22.1}$$

(Note that q is the total contact stress between the soil and the foundation.) Many simple foundations, including piles, are constructed from solid concrete which has unit weight γ_c only a little larger than that of soil, so W ($\approx \gamma_c AD$) depends on the size of the foundation. Some foundations are hollow, particularly where they are used for parking cars, in which case the weight W is relatively small.

Outside the foundation the total vertical stress at depth D is $\sigma_z = p_0$, where

$$p_0 = \gamma D \tag{22.2}$$

The net bearing pressure q_n is the change of total vertical stress at the base of the foundation and is given by

$$q_n = q - p_0 \tag{22.3}$$

Notice that q_n could be either positive or negative depending on the magnitudes of F and W, both of which would be very small for an underground car park or a submerged tank. If q_n is positive the foundation will settle, but if it is negative (i.e. the total stress at foundation level reduces) the foundation will rise. By careful design of a compensated foundation it is possible to have $q_n \approx 0$ so that settlements are negligible.

22.2 FOUNDATION BEHAVIOUR

Figure 22.3(a) shows a simple shallow foundation with a bearing pressure q and a settlement ρ. If the foundation is rigid (e.g. concrete) the settlement ρ will be uniform and the bearing pressure will vary across the foundation. If, on the other hand, the foundation is flexible (e.g. an earth embankment) the bearing pressure will be uniform but the settlements will vary. Figure 22.3 illustrates mean values of q and ρ for each case. Figure 22.3(b) shows the relationship between q and ρ for either drained or undrained loading. As the bearing pressure increases the settlements start to accelerate and at some load q_c the foundation can be said to have failed

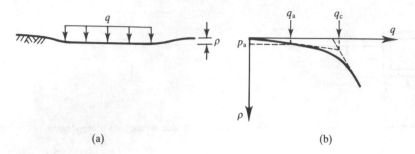

Figure 22.3 Loading and settlement of a foundation.

because the settlements have become very large. (Notice that as the foundation settles the bearing pressure continues to increase because the depth of the foundation increases.) The bearing pressure at failure is the bearing capacity of the foundation and it is the gross, not the net, bearing pressure.

Obviously you cannot load a building foundation close to its bearing capacity q_c as the settlements would then be too large and the building would probably be damaged (although it may not fall down). To limit the settlements to some allowable limit ρ_a it is necessary to reduce the bearing pressure to some allowable bearing pressure q_a, as shown in Fig. 22.3(b). In practice this is usually achieved by applying a factor of safety (or a load factor) to the bearing capacity (see Sec. 22.5).

Figure 22.4(a) shows the bearing pressure of a foundation increased to q_a slowly so that the loading is drained. The foundation settlements increase in parallel with the loading and terminate as ρ_d as shown in Fig. 22.4(b). Figure 22.4(c) shows the same loading increased quickly so the loading is undrained and there is an immediate, undrained settlement ρ_i, as shown in Fig. 22.4(d). The undrained loading raises the pore pressure in the soil below the foundation and dissipation of the excess pore pressures causes consolidation settlements to occur. The settlement at some

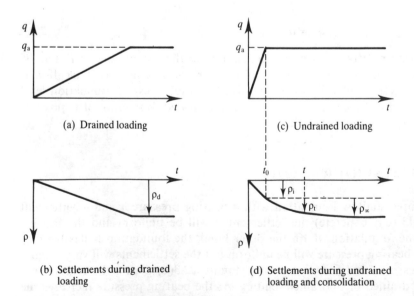

Figure 22.4 Loading and settlement of foundations.

time t after the start of consolidation is ρ_t and the final consolidation settlement which occurs after a relatively long time is ρ_∞. (Notice that the loadings and settlements shown in Fig. 22.4 are similar to those shown in Figs 6.9 and 6.10 which describe the fundamental differences between drained and undrained loading and consolidation.)

Generally, engineers designing foundations will need to calculate all, or some, of the following:

1. The bearing capacity q_c (to ensure that the foundation has an adequate margin of safety against collapse).
2. The allowable bearing pressure q_a and either the drained settlements ρ_d or the (undrained) immediate settlement ρ_i.
3. For consolidation after loading, the final consolidation settlement ρ_∞ and the variation of settlement ρ_t with time.

22.3 STRESS CHANGES IN FOUNDATIONS

The changes of stress and water content during undrained loading and subsequent consolidation of a foundation are illustrated in Fig. 22.5. In Fig. 22.5(a) the total stresses on a typical element below the foundation are τ and σ and the pore pressure is illustrated by the rise of water in a standpipe. In Fig. 22.5(b) the total stress path A → B corresponds to increases of σ and τ due to the loading of the foundation. The effective stress path is A′ → B′, which corresponds to undrained loading with constant water content, as shown in Fig. 22.5(c). The exact effective stress path A′ → B′ will depend on the characteristics of the soil and its initial overconsolidation ratio, as discussed in Chapter 11.

As shown in Fig. 22.5(b), the pore pressure immediately after construction u_i is greater than

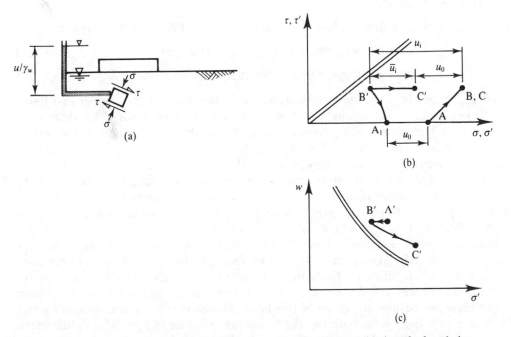

Figure 22.5 Changes of total and effective stress during loading and consolidation of a foundation.

the steady pore pressure u_0 and so the initial excess pore pressure $\bar{u}_i$ is positive. As time passes the total stresses remain essentially unchanged at B, since the foundation loading does not change, but the pore pressures drop. The effective stress path is B' → C', which corresponds to compression and an increase in the mean normal effective stress, as shown in Fig. 22.5(b) and (c).

The foundation can be said to fail if all the elements along a critical slip surface reach the critical state line. The distance of B' from the critical state line is a measure of the factor of safety of the foundation and Fig. 22.5 demonstrates that the factor of safety of a foundation generally increases with time but there will be continuing settlements due to consolidation.

22.4 BEARING CAPACITY OF SHALLOW FOUNDATIONS

The bearing capacity of a foundation can be calculated using the upper and lower bound methods (Chapter 18) or the limit equilibrium method (Chapter 19).

(a) Undrained Bearing Capacity

The bearing capacity of the simple shallow foundation shown in Fig. 22.6(a) is given by

$$q_c = s_u N_c + p_0 \tag{22.4}$$

or

$$F_c + W = s_u N_c B + \gamma D B \tag{22.5}$$

where N_c is a bearing capacity factor. For a long rectangular foundation at the ground surface identical upper and lower bounds obtained in Sec. 18.6 (see Eqs 18.43 and 18.46) are equivalent to $N_c = 2 + \pi$. The bearing capacity factor N_c depends only on the shape and depth of the foundation and values given by Skempton (1951) are shown in Fig. 22.6(b).

(b) Drained Bearing Capacity

The bearing capacity of the simple shallow foundation shown in Fig. 22.7(a) is given by

$$q_c = \tfrac{1}{2}(\gamma - \gamma_w)BN_\gamma + (\gamma - \gamma_w)(N_q - 1)D + \gamma D \tag{22.6}$$

or

$$F_c + W = \tfrac{1}{2}(\gamma - \gamma_w)B^2 N_\gamma + (\gamma - \gamma_w)(N_q - 1)BD + \gamma BD \tag{22.7}$$

where N_γ and N_q are bearing capacity factors. These could be obtained from upper and lower bound or limit equilibrium calculations, but these are lengthy so I have not given them here. The bearing capacity factors N_γ and N_q depend principally on the friction angle ϕ' and values given by Terzaghi (1943) are shown in Fig. 22.7(b). Notice that terms such as $(\gamma - \gamma_w)B$ or $(\gamma - \gamma_w)D$ represent effective stresses at depths B and D respectively in cases where the water table is at ground level; these effective stresses largely govern the soil strength. Equations (22.6) and (22.7) apply when the water table is at the ground surface; if the water table is at the base of the foundation put $\gamma_w = 0$ into the term containing N_q and if the water table is very deep and below the influence of the foundation put $\gamma_w = 0$ throughout.

These bearing capacity equations (Eqs 22.4 to 22.7), together with the bearing capacity factors N_c, N_q and N_γ given by Skempton (1951) and by Terzaghi (1943), can be used to calculate the ultimate bearing capacity of simple foundations. Other tables and charts for bearing capacity factors for deep foundations and for foundations with eccentric or inclined loads have been published, but these are beyond the scope of this book. Notice that if the soil becomes water so that $s_u = 0$ or $\phi' = 0$ and $\gamma = \gamma_w$, both Eqs (22.5) and (22.7) reduce to $F_c + W = \gamma_w BD$, which is a statement of Archimedes' principle.

(a)

$$\frac{N_c \text{ (rectangular)}}{N_c \text{ (square)}} = 0.84 + 0.16 \frac{B}{L}$$

D/B

(b) (*After* Skempton, 1951)

Figure 22.6 Bearing capacity factors for undrained loading of foundations.

22.5 CHOICE OF SOIL STRENGTH AND LOAD FACTOR FOR FOUNDATIONS

For slope stability (see Chapter 20) and for overall stability of retaining walls (see Chapter 21) the calculations may be carried out using the critical state strength with factors of safety that reflect the uncertainties in the determinations of the soil parameters and pore pressures. This technique is suitable for soil structures where the principal criterion for design is the ultimate stability, but it is unsuitable for foundations and other structures where the principal criterion for design is the magnitude of the settlements or ground movements. The principal problem is that the ratio of stiffness (which controls ground movements) to strength (which controls ultimate failure) is not a constant, even for a particular soil, so there is no constant ratio between bearing capacity and allowable bearing pressure.

Figure 22.8(a) shows typical stress–strain curves for samples of the same soil on the wet side of critical (i.e. loose or lightly overconsolidated) or on the dry side of critical (i.e. dense or heavily overconsolidated). These have the same critical state strength ϕ'_c but very different shear stiffnesses G' (even allowing for non-linear behaviour) and the sample on the dry side of critical has a peak. The corresponding load–settlement curves for the same foundation are shown in Fig. 22.8(b). Since the soil has a unique value of ϕ'_c we would calculate the same value of q_c

(a)

(b) (*After* Terzaghi, 1943)

Figure 22.7 Bearing capacity factors for drained loading of foundations.

(a)

(b)

Figure 22.8 Settlement of foundations on soils on the wet side and on the dry side of critical.

from the critical state strength, even though the ultimate bearing capacities would probably be different. However, because of the very different stiffnesses of the soil at states on the wet side and on the dry side of critical the allowable bearing pressures q_d and q_w would be very different to achieve the same settlement ρ_a; this means that a factor of safety to limit settlements based on the critical state strength ϕ'_c would also have to be very different. Alternatively, the same factor of safety based on the peak strength ϕ'_p would lead to large bearing pressures for foundations on dense soil initially on the dry side of critical and smaller pressures for foundations on loose soil initially on the wet side of critical.

The factor used to reduce the bearing pressure to limit settlements is not really a factor of safety as it is not intended to account for uncertainties; it is better to call it a load factor. For a foundation the load factor should logically be applied to the net bearing pressure q_n as it is this that causes settlement or heave. Notice that with a compensated foundation for which $q_n = 0$ there is no need to apply any load factor as the settlements and heave will be zero in any case. There is no real hard and fast rule for selecting an appropriate load factor for foundation designs; in practice the factors usually used are in the region 2 to 3. In any case the settlements of the foundation should be calculated independently using the methods described in Secs 22.7 and 22.8.

22.6 FOUNDATIONS ON SAND

Foundations on sand will be drained and the settlements ρ_d will occur as the loads are applied, as shown in Fig. 22.4(b). Figure 22.8 also illustrates the different behaviour of a foundation on a dense sand initially on the dry side of critical and a loose sand initially on the wet side of critical and shows that for a given allowable settlement ρ_a the allowable bearing pressures q_d and q_w depend on the initial relative density. A simple and logical design procedure would be to relate the allowable bearing pressure directly to the relative density (or the density of the initial state from the critical state line) measured in some suitable *in situ* test.

The routine test to measure relative density is the standard penetration test (SPT) described in Sec. 16.5. The result is given as a blowcount value N, which varies from small values (1 to 5) when the soil is at its loosest state to large values (over 50) when the soil is at its densest state. A simple relationship between the SPT-N value and the allowable bearing pressure was given by Terzaghi and Peck (1967) and a simple rule of thumb is

$$q_a = 10N \text{ kPa} \tag{22.8}$$

This bearing pressure will give settlements of the order of 25 mm (1 inch). Because at relatively small loads the load settlement curve in Fig. 22.8(b) is approximately linear, halving the bearing pressure will give about half the settlement and so on.

22.7 FOUNDATIONS ON ELASTIC SOIL

An assumption commonly made in practice is that soil is linear and elastic and there are a number of standard solutions for distributions of stresses and ground movements around foundations subjected to a variety of loads. These solutions have generally been obtained by integrating solutions for point loads and so they employ the principle of superposition which is valid only for linear materials. We have seen earlier (Chapters 12 and 13) that soils are usually neither elastic nor linear and so these solutions are not strictly valid, although the errors in calculation of stresses are likely to be considerably less than those in the calculation of ground movement.

The changes of the vertical stress $\delta\sigma_z$ and the settlements $\delta\rho$ at a point in an elastic soil due to a change δQ of a point load at the surface, shown in Fig. 22.9, are given by

$$\delta\sigma_z = \frac{3\delta Q}{2\pi R^2}\left(\frac{z}{R}\right)^3 \tag{22.9}$$

$$\delta\rho = \frac{\delta Q(1+v)}{2\pi ER}\left[\left(\frac{z}{R}\right)^2 + 2(1-v)\right] \tag{22.10}$$

where E and v are Young's modulus and Poisson's ratio. Although these expressions lead to

Figure 22.9 Stresses and settlements due to a point load.

infinite stresses and settlements immediately below the point load where $z = R = 0$, they can be used to calculate stresses and settlements some way below small foundations.

For circular or rectangular foundations on elastic soil the changes of vertical stress $\delta\sigma_z$ and settlement $\delta\rho$ at a point below a foundation due to a change of bearing pressure δq are given by

$$\delta\sigma_z = \delta q I_\sigma \tag{22.11}$$

$$\delta\rho = \delta q B \frac{1 - v^2}{E} I_\rho \tag{22.12}$$

where I_σ and I_ρ are dimensionless influence factors and B is the width or the diameter of the foundation. The values for the influence factors depend principally on the geometry of the foundation and, to a lesser extent, on the value of Poisson's ratio. Notice Eqs (22.9) and (22.11) do not contain either E or v and so the vertical stress in elastic soil depends only on the shape and loading of the foundation. A comprehensive set of tables and charts for influence factors for a wide variety of loading cases are given by Poulos and Davis (1974). Values for the most common simple cases for circular and rectangular loaded areas are shown in Figs 22.10 and 22.11.

Figure 22.10 Influence factors for stresses and settlements below the centre of a circular foundation.

(a)

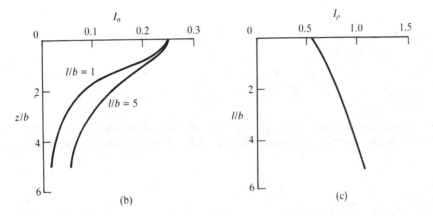

(b)

(c)

Figure 22.11 Influence factors for stresses and settlements below the corner of a rectangular foundation.

Figure 22.12 Division of rectangular loaded areas.

To determine values inside or outside a rectangular or irregularly shaped area you can simply divide the region into a number of rectangles, determine $\delta\sigma_z$ or $\delta\rho$ at the corners of the various rectangles and, making use of the principle of superposition, add or subtract the individual effects. For example, for the L-shaped building in Fig. 22.12 the stresses and settlements at the corner E can be found by adding the effects of the rectangles DABE, BCFE and HGDE; the stresses and settlements at the external point J can be found by subtracting the effects of the rectangle HEFJ from those of the rectangle GACJ.

In selecting values for Young's modulus and Poisson's ratio it is necessary to distinguish between drained and undrained loading. For drained loading choose the parameters E' and v' corresponding to effective stresses and for undrained loading choose E_u and $v_u = 0.5$ corresponding to undrained, constant volume loading. The basic relationship between the elastic shear modulus G and the elastic bulk modulus E (see Sec. 3.4) is

$$G = \frac{E}{2(1 + v)} \tag{22.13}$$

For an elastic material for which shear and volumetric effects are decoupled we have $G' = G_u$ and hence

$$\frac{E'}{2(1 + v')} = \frac{E_u}{2(1 + v_u)} \tag{22.14}$$

or, with $v_u = 0.5$,

$$E_u = \frac{3E'}{2(1 + v')} \tag{22.15}$$

The settlements of a foundation for drained loading ρ_d or for undrained loading ρ_u are given by Eqs (22.10) or (22.12) with the appropriate values for E and v. Hence, making use of Eq. (22.15),

$$\frac{\rho_u}{\rho_d} = \frac{3E'}{4(1 - v'^2)E_u} = \frac{1}{2(1 - v')} \tag{22.16}$$

and, taking a typical value of $v' = 0.25$, we have $\rho_u = 0.67\rho_d$. Thus, for foundations on an infinitely deep bed of elastic soil the settlements for undrained loading are of the order of two-thirds those for drained loading of the same foundation; the difference is made up by the additional settlements that occur due to consolidation after undrained loading. If the depth of the soil is relatively small compared to the width of the foundation so that the conditions in the soil are one-dimensional (see Sec. 22.8), $\rho_u = 0$.

These analyses, based on the theories of linear elasticity, are interesting and informative but we must not forget that soils are highly non-linear and inelastic over most of the range of loading of practical importance and the results should be viewed accordingly.

22.8 SETTLEMENTS FOR ONE-DIMENSIONAL LOADING

An assumption commonly made is that the thickness of a compressible soil layer is small compared to the width of the loaded foundation; so that the horizontal strains can be neglected. In this case the conditions of stress, strain and consolidation in the ground, shown in Fig. 22.13(a), are the same as those in the one-dimensional oedometer test described in Sec. 7.6 and 8.4 and shown in Fig. 22.13(b).

In the oedometer test the vertical strains $\delta\varepsilon_z$ are given by Eq. (8.9) as

$$\delta\varepsilon_z = m_v \, \delta\sigma'_z \tag{22.17}$$

where, for complete consolidation when $\bar{u} = 0$, we have $\delta\sigma'_z = \delta\sigma_z$. At the ground the surface

(b) Oedometer test

(a) Wide foundation

Figure 22.13 One-dimensional consolidation in foundations.

settlements due to consolidation $\delta\rho_c$ are given by

$$\frac{\delta\rho_c}{z} = \delta\varepsilon_z = m_v\,\delta\sigma'_z \tag{22.18}$$

where, for complete consolidation, we have $\delta\sigma'_z = \delta q$, where δq is the change of bearing pressure at the surface (i.e. the net bearing pressure). Final consolidation settlements for wide foundations can be calculated using Eq. (22.18). However, because the one-dimensional compression and swelling behaviour of soil is non-linear m_v is not a soil constant and it is necessary to measure m_v in an oedometer test in which the initial stress and the change of stress both correspond to those in the ground.

The rate at which consolidation settlements occur in one-dimensional oedometer tests was considered in Chapter 14. General solutions for rates of consolidation emerge as relationships between the degree of consolidation U_t and the time factor T_v. These are defined as

$$U_t = \frac{\Delta\rho_t}{\Delta\rho_\infty} \tag{22.19}$$

$$T_v = \frac{c_v t}{H^2} \tag{22.20}$$

where $\Delta\rho_t$ and $\Delta\rho_\infty$ are the settlements at times t and $t = \infty$, c_v is the coefficient of consolidation and H is the drainage path length.

Relationships between U_t and T_v depend on the geometry of the consolidating layer and its drainage conditions and on the distribution of initial excess pore pressure. The most common drainage conditions are one-dimensional or radial, as shown in Fig. 22.14. For one-dimensional drainage the seepage may be one-way towards a drainage layer at the surface, two-way towards drainage layers at the base and at the surface or many-way towards silt or sand layers distributed through the deposit. For radial drainage seepage is towards vertical drains placed on a regular grid. In each case the drainage path length, H or R, is the maximum distance travelled by a drop of water seeping towards a drain.

For one-dimensional consolidation the relationships between U_t and T_v for different initial excess pore pressure conditions are given in Fig. 14.10 in terms of T_v. These could also be given in terms of T_v plotted to a logarithmic scale, as shown in Fig. 22.15(a), which corresponds to consolidation with the initial excess pore pressure $\bar{u}_i$ uniform with depth. Figure 22.15(b) is for

(a) One-way drainage

(b) Two-way drainage

(c) Drainage to horizontal sand layers

(d) Radial drainage to vertical drains

Figure 22.14 Drainage conditions in foundations.

radial consolidation where

$$T_r = \frac{c_r t}{R^2} \tag{22.21}$$

$$n = \frac{R}{r} \tag{22.22}$$

As discussed in Chapter 14, these can be used to calculate either the settlement after a given time or the time for a given settlement. Although, in theory, complete consolidation will require infinite time a reasonable approximation is that T_v or $T_r \approx 1.0$ at $U_t = 1.0$.

22.9 SUMMARY

1. Foundations transmit loads to the ground. As the load increases the foundation settles and it fails when the settlements become very large. Foundations may be shallow or they may be deep to take advantage of the general increase of strength and stiffness of soils with depth.
2. The bearing pressure q is the contact stress between the foundation and the soil. The net bearing pressure of a deep foundation is the change of bearing pressure; this may be positive so the foundation settles or it may be negative so it heaves. The bearing pressure q and the net bearing pressure q_n are given by

$$q = \frac{F + W}{A} \tag{22.1}$$

$$q_n = q - p_0 \tag{22.3}$$

3. Under a foundation pore pressures generally increase with undrained loading and, with time,

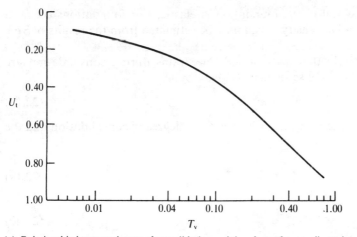

(a) Relationship between degree of consolidation and time factor for one-dimensional consolidation. (*After* Taylor, 1948.)

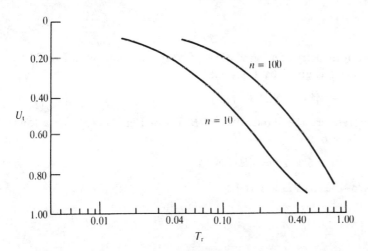

(b) Relationship between degree of consolidation and time factor for radial consolidation. (*After* Barron, 1948.)

Figure 22.15 Solutions for rate of consolidation.

these dissipate as the soil consolidates. As a result further settlements occur but effective stresses and safety factors increase.

4. The bearing pressure when the foundation fails is the bearing capacity q_c given by

$$q_c = s_u N_c + p_0 \tag{22.4}$$

$$q_c = \tfrac{1}{2}(\gamma - \gamma_w)BN_\gamma + (\gamma - \gamma_w)(N_q - 1)D + \gamma D \tag{22.6}$$

for undrained and drained loading respectively, where N_c, N_γ and N_q are bearing capacity factors.

5. An important criterion for foundation design is the need to limit the settlements. This may be done by applying a load factor to the net bearing pressure. Alternatively, settlements may

be calculated assuming that the soil in the foundation is elastic. For foundations on sand settlements are related to the relative density which may be estimated from the results of SPT tests.

6. For wide foundations on relatively thin beds of soil the strains during consolidation are one-dimensional. The magnitude of the settlement is given by

$$\delta\rho_c = z m_v \, \delta\sigma'_z \tag{22.18}$$

The rate of settlement is given by the relationship between the degree of consolidation and the time factor, which are given by

$$U_t = \frac{\Delta\rho_t}{\Delta\rho_\infty} \tag{22.19}$$

$$T_v = \frac{c_v t}{H^2} \tag{22.20}$$

A reasonable approximation is $T_v = 1$ when $U_t = 1$.

WORKED EXAMPLES

Example 22.1: Undrained bearing capacity of a foundation   For the foundation in Fig. 22.16 the ultimate load for undrained loading is given by Eq. (22.5):

$$F_c + W = s_u N_c B + \gamma D B$$

If the unit weights of soil and concrete are the same, $W = \gamma D B$. From Fig. 22.6(b), for a long foundation with $D/B \approx 1$ we have $N_c = 6$ and

$$F_c = 30 \times 6 \times 2.5 = 450 \text{ kN/m}$$

If the applied load is $F_a = 300$ kN/m the load factor is 1.5.

300 kN/m

$s_u = 30\,\text{kPa}$
$\phi' = 25°$
$\gamma = 20\,\text{kN/m}^3$

2 m

2.5 m

Figure 22.16

Example 22.2: Drained bearing capacity of a foundation   For the foundation in Fig. 22.16 the ultimate load for drained loading is given by Eq. (22.7):

$$F_c + W = \tfrac{1}{2}(\gamma - \gamma_w)B^2 N_\gamma + (\gamma - \gamma_w)(N_q - 1)BD + \gamma BD$$

As before, $W = \gamma BD$. From Fig. 22.7(b), for $\phi' = 25$, $N_\gamma = 8$ and $N_q = 11$ and

$$F_c = \tfrac{1}{2}(20 - 10)2.5^2 \times 8 + (20 - 10)(11 - 1)2.5 \times 2 = 750 \text{ kN/m}$$

If the applied load is $F_a = 300$ kN/m the load factor is 2.5.

Example 22.3: Settlements of an embankment The embankment in Fig. 22.17 is sufficiently wide so that the strains and seepage in the soil can be assumed to be one-dimensional. From Eq. (22.18) the magnitude of the final consolidation settlement is

$$\rho_c = m_v z \, \Delta\sigma'_z = 5 \times 10^{-4} \times 8 \times 100 = 0.40 \text{ m}$$

(a) From Fig. 22.15(a) the time when the settlement is complete (i.e. when $U_t = 1.0$) corresponds to $T_v = 1.0$. Hence, from Eq. (22.20),

$$t = \frac{T_v H^2}{c_v} = \frac{1.0 \times 8^2}{2} = 32 \text{ years}$$

(b) After 5 years the time factor is

$$T_v = \frac{c_v t}{H^2} = \frac{2 \times 5}{8^2} = 0.16$$

From Fig. 22.15(a) this corresponds to a degree of consolidation $U_t \approx 0.50$ and the settlement after 5 years is

$$\rho_t = U_t \rho_\infty = 0.50 \times 0.40 = 0.20 \text{ m}$$

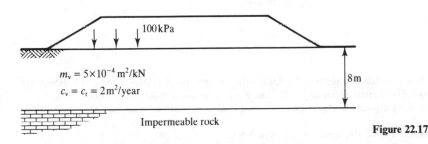

100 kPa

$m_v = 5 \times 10^{-4} \text{ m}^2/\text{kN}$

$c_v = c_r = 2 \text{ m}^2/\text{year}$

8 m

Impermeable rock

Figure 22.17

Example 22.4: Settlements with drains In order to speed up the settlements of the embankment in Example 22.3 sand drains are installed in the clay. The drains are 200 mm in diameter ($r = 100$ mm) and they are spaced 2 m apart ($R = 1.0$ m).

From Fig. 22.15(b), with $n = R/r = 10$, the time when settlement is complete (i.e. when $U_t = 1.0$) corresponds to $T_v = 1.0$. Hence, from Eq. (22.21),

$$t = \frac{T_r R^2}{c_r} = \frac{1.0 \times 1.0}{2} = 0.5 \text{ years}$$

Example 22.5: Calculation of stresses and settlements in elastic soil Figure 22.18 shows a circular water tank at the surface of a deep bed of elastic soil. For $\delta q = 5 \times 10 = 50$ kPa the changes of vertical stress and the settlements for drained loading are given by Eqs (22.11) and (22.12):

$$\delta\sigma'_z = \delta q I_\sigma = 50 I_\sigma \quad \text{kPa}$$

$$\delta\rho = \delta q B \frac{1 - v'^2}{E'} I_\rho = 50 \times 10 \frac{(1 - 0.25^2) \times 10^3}{10 \times 10^3} I_\rho = 47 I_\rho \quad \text{mm}$$

where I_σ and I_ρ are given in Fig. 22.10.

Figure 22.18

(a) At point A, $z/a = 0$ so $I_\sigma = 1.0$ and $I_\rho = 1.0$; hence

$$\delta\sigma'_z = 50 \text{ kPa}$$

$$\delta\rho = 47 \text{ mm}$$

(b) At point B, $z/a = 1$ so $I_\sigma = 0.65$ and, for $v' = 0.25$ (interpolating between the data for $v' = 0$ and $v' = 0.5$), $I_\rho = 0.65$; hence

$$\delta\sigma'_z = 33 \text{ kPa}$$

$$\delta\rho = 31 \text{ mm}$$

REFERENCES

Barron, R. A. (1948) 'Consolidation of fine grained soils by drain wells', *Trans. Am. Soc. Civil Engng*, **113**, 718–754.

Poulos, H. G. and E. H. Davis (1974) *Elastic Solutions for Soil and Rock Mechanics*, Wiley, New York.

Skempton, A. W. (1951) 'The bearing capacity of clays', *Proceedings of Building Research Congress*, Vol. 1, pp. 180–189, ICE, London.

Taylor, D. W. (1948) *Fundamentals of Soil Mechanics*, Wiley, New York.

Terzaghi, K. (1943) *Theoretical Soil Mechanics*, Wiley, New York.

Terzaghi, K. and R. B. Peck (1967) *Soil Mechanics in Engineering Practice*, Wiley, New York.

FURTHER READING

Atkinson, J. H. (1981) *Foundations and Slopes*, McGraw-Hill, London.

Bowles, J. E. (1989) *Foundation Analysis and Design*, McGraw-Hill, New York.

BS 8004 (1986) *Code of Practice for Foundations*, British Standards Institution, London.

Burland, J. B. and C. P. Wroth (1975) 'Settlement of buildings and associated damage', *Proceedings of Conference on Settlement of Structures*, Pentech Press, London.

Padfield, C. J. and M. J. Sharrock (1983) *Settlement of Structures on Clay Soils*, CIRIA, Special Publication 27, London.

Peck, R. B., W. E. Hanson and T. H. Thorburn (1974) *Foundation Engineering*, Wiley, New York.

TWENTY-THREE

PILED FOUNDATIONS

23.1 TYPES OF PILED FOUNDATIONS

Piles are long slender columns installed into the ground, often in groups. The principal purpose of piling is to transfer loads to stronger and stiffer soil or rock at depth, to increase the effective size of a foundation and to resist horizontal loads. Typically piles are made from steel or reinforced concrete and possibly timber. They may be driven or pushed into the ground or concrete piles may be cast *in situ* by pouring concrete into a drilled hole.

Some typical pile types are illustrated in Fig. 23.1. Figure 23.1(a) shows an end bearing pile where most of resistance is developed at the toe and Fig. 23.1(b) shows a friction pile where a significant contribution to the pile capacity is developed by shear stresses along the sides. Figure 23.1(c) shows raking piles to resist horizontal loads and Fig. 23.1(d) is a pile group joined at the top by a pile cap. Notice that the pile on the left in Fig. 23.1(c) is in tension and so all the resistance comes from shear stress on the sides of the pile.

Figure 23.2 shows the loads on a single pile: the applied load Q is resisted by a force at the base Q_b and a force Q_s due to the shear stresses between the soil and the pile shaft; hence

$$Q = Q_s + Q_b \tag{23.1}$$

In conventonal pile analysis the weight of the pile is taken to be the same as the weight of soil displaced by the pile and both are neglected. In any case these forces are usually small compared with the applied loads, which are typically in the range 500 to 5000 kN and may be considerably larger. Figure 23.2(b) illustrates the increase in base resistance and shaft friction with displacement. The shaft friction increases more quickly than the base resistance and reaches an ultimate state at relatively small strains.

Piles or pile groups may be loaded drained or undrained and the basic total and effective stress paths will be similar to those for shallow foundations, shown in Fig. 22.5. Generally, piles installed in a clay soil will settle with time as the excess pore pressures generated by undrained loading dissipate and the effective stresses and strength of the soil increases. There may, however, be stress changes caused by installation which would cause swelling and softening of the soil around a bored and cast *in situ* pile or compression and consolidation around a driven pile.

Figure 23.1 Types of piled foundations.

(a) Base resistance and shaft friction

(b) Load settlement curves

Figure 23.2 Pile resistance.

23.2 BASE RESISTANCE OF SINGLE PILES

The base resistance of a single pile is given by

$$Q_b = q_b A_b \tag{23.2}$$

where q_b is the bearing capacity at the toe and A_b is the area of the pile base. The general principles for calculation of the bearing capacity of piles are similar to those for shallow foundations described in Chapter 22. The mechanism of slip surfaces at the tip of a pile appropriate for an upper bound or limit equilibrium calculation will be similar to that shown in Fig. 23.3 and we would expect the bearing capacity factors for piles to be larger than those for shallow foundations. For undrained loading the bearing capacity is given by

$$q_b = s_u N_c \tag{23.3}$$

(a)

(From Berezantzev, Khristoforov and Golubkov, 1961)

(b)

Figure 23.3 Base resistance of piles.

and, for square or circular piles, $N_c \approx 9$ (Skempton, 1951). For drained loading the bearing capacity is given by

$$q_b = \sigma'_z N_q \tag{23.4}$$

where σ'_z is the vertical effective stress at the level of the toe of the pile. Values for the bearing capacity factor N_q depend principally on ϕ' and there are a number of published relationships based on theory and experiment. The values shown in Fig. 23.3(b) are those given by Berezantzev, Khristoforov and Golubkov (1961).

The choice of the appropriate value of ϕ' is problematical. Soil below the toe of a driven pile will be highly strained during driving while there is the possibility of stress relief and softening at the base of a bored and cast *in situ* pile during construction. Consequently, in both cases a rational design method would take the critical friction angle ϕ'_c to determine a value of N_q for pile design. However, experiments and *in situ* tests indicate that use of ϕ'_c with the values of N_q in Fig. 23.3(b) leads to overconservative designs and often a peak friction angle ϕ'_p is used in practice.

The base resistance of a single pile may also be estimated from the *in situ* probing tests described in Chapter 16. The end bearing capacity of a pile is often equated with the cone resistance measured during a static cone test (sometimes with a correction for the different sizes of the pile and the cone) or derived from the standard penetration test N value.

23.3 SHAFT FRICTION ON PILES

From Fig. 23.4 resistance due to shaft friction on a circular pile, diameter D, is given by

$$Q_s = \pi D \int_0^L \tau_s \, dz \tag{23.5}$$

where τ_s is the shear stress mobilized between the pile and the soil. The value of τ_s is very difficult to determine; it depends on soil, on the pile material and particularly on the method of installation. For undrained loading of piles in clay,

$$\tau_s = \alpha s_u \tag{23.6}$$

Figure 23.4 Shaft resistance of piles.

where α must be in the range $0 \leqslant \alpha \leqslant 1$. Typically α is taken to be about 0.5 for both driven and cast *in situ* piles. For drained loading,

$$\tau'_s = \sigma'_h \tan \delta' = K\sigma'_z \tan \delta' \qquad (23.7)$$

where K is the ratio of the horizontal and vertical effective stresses σ'_h/σ'_z and must be in the range $K_a \leqslant K \leqslant K_p$ (where K_a and K_p are the active and passive earth pressure coefficients discussed in Chapter 21); δ' is the friction angle for shearing between the pile and the soil and for a rough pile this will be in the range $\phi'_r \leqslant \delta' \leqslant \phi'_p$. For clays, Eq. (23.7) is often simplified to

$$\tau'_s = \beta\sigma'_z \qquad (23.8)$$

where $\beta = K \tan \delta'$ is an empirical parameter that depends on the nature of the soil and on the method of pile installation.

Pile installation influences both δ' and K but differently. When a pile is driven into the ground there will be very large shear displacements between the pile and the soil, and in clays these displacements will probably be enough to reduce the soil strength to its residual value. However, pile driving is likely to increase the horizontal effective stresses which will tend to increase the shaft friction. On the other hand if a pile is driven into cemented soil, the horizontal stress after driving and the available shaft friction could be very small indeed. A cast *in situ* concrete pile is likely to have very rough sides and so the available shearing resistance will lie between the peak and the critical state strength of the soil. However, boring a hole in the ground to construct a cast *in situ* pile will reduce the horizontal stresses which may be reduced still further as the concrete shrinks during setting and curing. For both driven and cast *in situ* piles there are compensating effects on δ' and on K.

Notice that in a soil that is settling, perhaps due to the weight of fill placed at the surface or due to groundwater lowering, the shaft friction will act downwards on the pile as shown in Fig. 23.5, causing negative shaft friction.

23.4 PILE TESTING AND DRIVING FORMULAE

Because of the considerable uncertainties in the analysis of pile load capacity, both in calculation of base resistance and shaft friction, some of the piles on a job will often be subjected to load tests to demonstrate that their capacity is adequate. In typical tests loads will be applied in excess of the design working load and the deflections measured. The loads may be applied in stages and maintained at each increment (like in an oedometer test) or applied at a constant rate of penetration. The latter method is found to give more consistent results and better definition of failure loads.

Figure 23.5 Negative shaft friction due to ground settlement.

Figure 23.6 Pile driving formulae.

The capacity of a pile can be inferred from its resistance to driving. The basis of these so-called pile driving formulae is that the work done by the hammer (less any losses) is equal to the work done as the pile penetrates the ground. For the simple drop hammer weight W falling through h shown in Fig. 23.6 the pile capacity Q is related to the set s (i.e. the displacement) for a single blow by

$$Q_s = Wh \qquad (23.9)$$

Equation (23.9) is a very simple driving formula, too approximate to be used in practice, but it is the basis of other formulae which include terms to take account of energy losses in the hammer and in the pile.

23.5 CAPACITY OF PILE GROUPS

In a group of piles like that shown in Fig. 23.1(d), there will be interactions between neighbouring piles so that the capacity of each pile in the group will be reduced. A group efficiency η is given by

$$F = n\eta Q \qquad (23.10)$$

where F is the total load on the group, n is the number of piles in the group and Q is the capacity of an individual pile on its own. Values for the efficiency η decreases with reduced spacing of the piles, roughly as shown in Fig. 23.7(b).

If the pile spacing is relatively close, as shown in Fig. 23.7(c), it is more appropriate to consider the group as an equivalent foundation of base area A and depth L_g, where $L_g \approx \frac{2}{3}L$. The bearing capacity q_c of the block is calculated using the methods for shallow foundations described in Chapter 22 and the shear stresses on the sides of the block are calculated assuming that the ultimate shear stresses developed correspond to the strength of the soil.

23.6 SUMMARY

1. Piled foundations are used to lower the foundation into soil which is stiffer and stronger. The load capacity of a pile arises from base resistance and shaft friction.

(b)

(After Whitaker, 1970)

Figure 23.7 Capacity of pile groups.

2. Base resistance of a single pile is given by

$$q_b = s_u N_c \tag{23.3}$$

$$q_b = \sigma_v' N_q \tag{23.4}$$

for undrained and drained loading respectively.

3. The shaft friction of a single pile is given by

$$\tau_s = \alpha s_u \tag{23.6}$$

or

$$\tau_s' = \beta \sigma_z' \tag{23.8}$$

where α is a shaft friction factor for undrained loading and, for drained loading, $\beta = K \tan \delta'$.

4. In practice the capacity of piles is often determined from full-scale load tests or from pile driving formulae. The capacity of groups of piles can be found from the capacity of a single pile with an efficiency factor or from the geometry of an equivalent foundation.

REFERENCES

Berezantzev, V. G., V. S. Khristoforov and V. N. Golubkov (1961) 'Load bearing capacity and deformation of piled foundations', *Proceedings of 5th International SMFE Conference, Paris*, Vol. 2.

Skempton, A. W. (1951) 'The bearing capacity of clays', *Proceedings of Building Research Congress*, Vol. 1, pp. 180–189, ICE, London.

Whitaker, T. (1970) *The Design of Piled Foundations*, Pergamon Press, London.

FURTHER READING

Fleming, W. G. K., A. J. Weltman, M. F. Randolph and W. K. Elson (1985) *Piling Engineering*, Surrey University Press, London.
Whitaker, T. (1970) *The Design of Piled Foundations*, Pergamon Press, London.

TWENTY-FOUR

GEOTECHNICAL CENTRIFUGE MODELLING

24.1 MODELLING IN ENGINEERING

Engineers frequently use scale models in conjunction with theoretical analyses. For example, wind tunnel modelling is used routinely by engineers to study the flow of air past vehicles, aircraft and buildings. Hydraulic engineers frequently use models to study the flow of water in river channels, tidal flow in estuaries and wave loading on structures. Scale modelling is used most often when the theoretical solutions contain major simplifications and approximations or when numerical solutions are very lengthy, as is often the case in geotechnical engineering.

A geotechnical model might be tested when it would be too difficult, expensive or dangerous to build and test a full-scale structure. For example, it would be very difficult to test the response of a large earth-fill dam to earthquake loading and it would be very dangerous to examine the collapse of a tunnel heading during construction. Usually a model will be smaller than the prototype (or full-scale) structure that it represents.

The principles for modelling fluid flows are well established and so too are the principles for geotechnical modelling. To achieve correct scaling in geotechnical models the unit weight of the soil is increased by accelerating the model in a geotechnical centrifuge.

At present modelling is used less frequently in geotechnical engineering than in other branches of civil engineering but it is an important and valuable technique and one that you should know about. Detailed discussion of geotechnical centrifuge modelling is obviously beyond the scope of this book and what I want to do in this chapter is simply to set out the basic principles and to describe the principal purposes of modelling.

24.2 SCALING LAWS AND DIMENSIONAL ANALYSIS

Normally a model and the prototype will be geometrically similar so that all the linear dimensions in a model will be scaled equally but, for various reasons, it is impossible to construct a model that behaves exactly like a large prototype in all respects. (You have probably noticed that the waves made by a model sailing boat are different from the waves made by a full-sized yacht.) Instead, the model should have similarity with the prototype in the aspect of behaviour under examination. For example, in a wind tunnel model of an aircraft wing the relationships between lift, drag and velocity should be similar while in a river model the relationships between water depths and velocities should be similar, but neither model need look very much like the prototype it represents. On the other hand, a model built by an architect or a railway enthusiast should look like the real thing.

The rules that govern the conditions for similarity between models and prototypes are well known and the simplest method for establishing scaling laws is by dimensional analysis. The basic principle is that any particular phenomenon can be described by a dimensionless group of the principal variables. Models are said to be similar when the dimensionless group has the same value and then the particular phenomenon will be correctly scaled. Often these dimensionless groups have names and the most familiar of these are for modelling fluid flow (e.g. the Reynolds number).

24.3 SCALING GEOTECHNICAL MODELS

In constructing a geotechnical model the objectives might be to study collapse, ground movements, loads on buried structures, consolidation or some other phenomenon during a construction or loading sequence. In earlier chapters of this book I showed that soil behaviour is governed to a very major extent by the current effective stresses (this is a consequence of the fundamental frictional nature of soil behaviour). Consequently, the stresses at a point in a model should be the same as the stresses at the corresponding point in the prototype.

Figure 24.1(a) shows the vertical total stress at a depth z_p in a prototype construction in the ground and Fig. 24.1(b) shows a similar point at a depth z_m in a model with a scale factor n (i.e. all the linear dimensions in the model have been reduced n times). In the prototype the vertical stress is

$$\sigma_p = g\rho z_p \qquad (24.1)$$

where ρ is the density of the soil and $g = 9.81$ m/s^2 is the accleration due to Earth's gravity. If the model is placed in a centrifuge and accelerated to n times g the stress at a depth in the model $z_m = z_p/n$ is

$$\sigma_m = ng\rho z_m = \frac{ng\rho z_p}{n} \qquad (24.2)$$

and $\sigma_m = \sigma_p$. Since the stresses at equivalent depths are the same the soil properties will also be the same (provided that the stress history in the model and prototype are the same) and the behaviour of the soil in the model will represent the behaviour of the soil in the prototype. Notice that you cannot reproduce the correct prototype stresses by applying a uniform surcharge to the surface of the model as, in this case, the stresses in the model will be approximately constant with depth rather than increasing linearly with depth as in the ground.

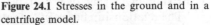

(a)   (b)

Figure 24.1 Stresses in the ground and in a centrifuge model.

Figure 24.2 Scaling for the stability of a model slope.

Another way of looking at the requirements of geotechnical modelling is through dimensional analysis. The stability of a slope for undrained loading was described in Sec. 20.8. For the prototype slope in Fig. 24.2(a) with height H_p and slope angle i (which is itself dimensionless), the stability depends on the undrained strength s_u, the height H_p and the unit weight $\gamma = g\rho$. These can be arranged into a dimensionless group

$$N_s = \frac{g\rho H}{s_u} \tag{24.3}$$

where N_s is a stability number. Notice that this is exactly the same as the stability number in Eq. (20.39). A model and a prototype are similar (i.e. they will both collapse in the same way) if they both have the same value of N_s. If the scale factor is n so that the model height H_m and the prototype height H_p are related by $H_m = H_p/n$ the stability numbers can be made equal by accelerating the model in a centrifuge to ng so that

$$N_s = \frac{g\rho H_p}{s_u} = \frac{ng\rho H_m}{s_u} \tag{24.4}$$

Thus the stability of the model slope illustrated in Fig. 24.2(b) will be the same as the stability of the prototype slope in Fig. 24.2(a) and if the slopes fail they will both fail in the same way.

The stresses, and the basic soil properties, in a prototype and in an nth scale model will be the same if the model is accelerated in a centrifuge to ng, but time effects may require a different scaling. There are several aspects of time in geotechnical engineering, the most important being associated with consolidation.

Consolidation due to dissipation of excess pore pressures with constant total stresses was discussed in Chapter 14. The rate at which excess pore pressures dissipate during one-dimensional consolidation is given by Eq. (14.34) and for similarity the time factor T_v in the model and in the prototype should be the same. From Eq. (14.25),

$$T_v = \frac{c_v t_p}{H_p^2} = \frac{c_v t_m}{H_m^2} \tag{24.5}$$

In a model with the same soil and pore fluid as the prototype, c_v is the same and if the scale is n we have $H_m = H_p/n$. Hence, from Eq. (24.5), the times for consolidation in the model and prototype are related by

$$t_m = \frac{t_p}{n^2} \tag{24.6}$$

so that consolidation will proceed much more rapidly in the model than in the prototype. For a typical scale factor $n = 100$, we have $t_m = 10^{-4} t_p$ so that 1 hour of model time represents approximately 1 year of prototype consolidation time.

The relationship between the rate at which excess pore pressures dissipate as drainage occurs and the rate of loading that generates additional excess pore pressures governs whether a particular construction event is drained, undrained or partly drained, as discussed in Sec. 6.8. Remember that for routine geotechnical calculations we have to assume either that the soil is fully drained or that it is fully undrained, in which case there will be subsequent consolidation. A model could, however, examine cases of partial drainage in which the rates of loading and consolidation were coupled.

If the accelerations in the prototype and in the model are related by the scale factor n and are given by

$$\frac{d^2 x_p}{dt_p^2} = a\omega^2 \sin(\omega t_p) \tag{24.7}$$

$$\frac{d^2 x_m}{dt_m^2} = na\omega^2 \sin(n\omega t_m) \tag{24.8}$$

then the displacements are given by

$$x_p = a \sin(\omega t_p) \tag{24.9}$$

$$x_m = \frac{a}{n} \sin(n\omega t_m) \tag{24.10}$$

and the times in the prototype and in the model are related by

$$t_p = nt_m \tag{24.11}$$

Any motion can be represented by a Fourier series which is a summation of sine functions and so the time scaling rule given by Eq. (24.11) applies to any displacement or loading. Notice that the scaling requirement for the rate of loading is that the times should be related by n, which is not the same as the requirement for modelling consolidation where the times should be related by n^2. Therefore it is not generally possible to model coupled loading and consolidation in the same model. This problem can be avoided by using a pore fluid such as silicon oil with a viscosity n times greater than that of water. In this case the coefficient of consolidation and the rate of consolidation in the model are deduced by n times so that the scaling $t_p = nt_m$ is then the same for both the rate of loading and the rate of consolidation.

24.4 PURPOSES OF MODELLING

It would be very convenient to be able to construct and test a scale model that reproduced all the significant features of the behaviour of a proposed construction. Unfortunately, however, this is not generally possible for a variety of reasons. The principal difficulties are rather like those associated with ground investigations and laboratory testing (see Chapters 7 and 16) and are due to test samples not being fully representative of the soil in the ground. It is also difficult to model geological history and complex construction sequences. Instead, geotechnical models are usually constructed and tested to meet specific objectives.

The principal purposes and categories of geotechical modelling were discussed by Taylor (1987) and these are as follows.

(a) Mechanistic Studies

The basic methodology of engineering design is that engineers imagine all the possible ways in which a proposed construction may fail or distort and they then carry out analyses that demonstrate that it will perform satisfactorily in any of these ways. Sometimes major failures occur when the construction finds some other way to fail or distort. For example, in the upper bound and limit equilibrium methods described in Chapters 18 and 19 it is necessary to define compatible mechanisms and the solutions depend on the mechanisms chosen. For relatively simple cases it is usually possible to choose the critical mechanisms from previous experience, but in novel and complex cases they may not be so obvious. In these cases relatively simple model tests may be carried out simply to observe qualitatively the way in which the structure distorts and fails, thus indicating the most appropriate analyses.

(b) Validation of Numerical Analyses

Design of geotechnical structures often requires complex numerical analyses using finite element, or similar, methods with non-linear and inelastic soil behaviour (see Chapter 13). These analyses are highly complex and before they are applied in design studies they should be tested against exact analytical solutions or against observations of the real events. Observations from relatively simple model tests can be used to test numerical analyses. The models should be similar to the proposed construction but, since the models are used only to calibrate the analyses, they need not reproduce all the details of the prototype.

(c) Parametric Studies

Another important procedure in design studies involves examining alternative construction details and investigating the consequences of different design assumptions. Furthermore, standard design codes and charts rely on studies of many different alternatives. Normally parametric studies are carried out using analytical or numerical methods, but model studies have a role to play in parametric studies, either on their own or together with other methods.

(d) Site-Specific Studies

In this case the model is intended to represent a particular construction so that the behaviour of the model is used directly to assess the behaviour of the prototype. It is obviously not easy to model all the details of the ground conditions and the construction and loading sequence; these are the most difficult type of centrifuge models to construct and test satisfactorily.

Model studies may be carried out for more than one purpose, for example combining validation of analyses with parametric studies. In practice, designs are vary rarely completed on the basis of model tests alone and model tests are almost always used in conjunction with numerical analysis.

24.5 GEOTECHNICAL CENTRIFUGES

In a geotechnical centrifuge, a model in a strong container is rotated in a horizontal plane about a vertical axis as shown in Fig. 24.3. At the model the centrifugal acceleration a is

$$a = ng = \omega^2 r \tag{24.12}$$

where r is the radius and ω is the angular velocity (in radians per second). To maintain a reasonably constant acceleration field through the model the radius r should be large compared with the size of the model.

The essential features of a geotechnical centrifuge are illustrated in Fig. 24.4. The motor

Figure 24.3 Centrifuge acceleration.

Figure 24.4 Characteristic features of a typical geotechnical centrifuge.

drives a vertical shaft at constant speed. The arm has an adjustable counterweight for balance and the model sits on a swing. At rest the swing hangs down, but as the arm rotates it swings up to a nearly horizontal position as shown. The purpose of the swing is so that the self-weight of the model always acts towards the base of the container; if you put a strong bucket containing water on the swing and start the centrifuge the water will remain level in the bucket.

The selection of the dimensions and speed for design of a geotechnical centrifuge is a matter of optimization between a number of conflicting requirements. A given prototype size could be represented by a small model tested at high accelerations or by a larger model at smaller accelerations; a given acceleration, or scale factor, can be achieved by a high-speed machine with a relatively small radius or by a machine with a larger radius rotating more slowly.

From Eq. (24.12) the acceleration is given by $\omega^2 r$, so a small-radius, high-speed machine is more efficient than one with a larger radius and lower speeds. If, however, the radius is not large compared to the depth of the model there may be significant variations of acceleration with depth in the model. A small model, requiring large accelerations, will be relatively easy to manufacture and handle, but it will be possible to install only a limited number of instruments. On the other hand, a larger model which can be more easily instrumented will be heavy and more difficult to manufacture and handle on to the centrifuge.

The mass of the model, including the soil, the strong container and all the ancillary equipment for loading and observing the model, is called the payload. The capacity of a centrifuge is often quoted as the product (in g-tonnes) of the maximum acceleration (i.e. the scale factor) and the maximum payload at that acceleration.

There is a very great variation in the dimensions and capacities of geotechnical centrifuges and some examples to illustrate cases throughout the range are given in Table 24.1. The optimization of size and capacity is determined largely by the resources of manpower available to the group who will run the facility, so that university groups tend to acquire machines requiring smaller and more easily managed models while commercial and government-run facilities tend to have machines able to test larger models that can accommodate more instrumentation.

As a very rough guide, about 50 per cent of the payload could be soil, with the remainder required for the strong container and other equipment. For the Acutronic 661 machine at City University the maximum payload of 400 kg could have about 200 kg of soil and this could be in a model (say) of 600 mm × 400 mm × 400 mm. At a scale factor of $n = 100$ (i.e. at an acceleration of 100 g) this represents a prototype volume of soil of 60 m × 40 m × 40 m or a 20 m thick plane strain slice 100 m wide and 50 m deep. A package up to 400 kg can be handled reasonably easily without expensive cranes and the Acutronic 661 machine represents an optimum size for a university facility.

Table 24.1

Centrifuge manufacturer	Location	R (m)	Payload (kg)	Acceleration (g)	Capacity (g tonnes)
MSE Mistral	City University	0.2	0.75	1500	1
Acutronic model 661	City University	1.8	400 200	100 200	40
Acutronic model 680	LCPC, Nantes	5.5	2200 1100	100 200	220
Krupp	Ruhr-University Bochum	4.1	2000	250	500

In the very small MSE Mistral centrifuge at City University the maximum size of the model is only about 80 mm × 80 mm × 20 mm, but at a scale factor of $n = 1000$ this represents a prototype volume of soil 80 m × 80 m × 20 m thick. In such a small model there is little opportunity for instrumentation, but the deformations of the model can be photographed or recorded on videotape using a stroboscopic flash. Very small machines are used principally for teaching and for student projects.

24.6 CONTROL AND INSTRUMENTATION IN CENTRIFUGE MODELS

During a typical geotechnical centrifuge model test the machine will be run at constant speed (i.e. at constant scale factor) while the model is loaded or unloaded and the behaviour observed. The requirements for control of loading and measurement of load and displacement in a model are broadly similar to those for laboratory tests described in Chapter 7.

Communication with the rotating model is through slip rings, as shown in Fig. 24.4. These may transmit fluids (e.g. water, gas or hydraulic oil) or power to operate motors or valves, and they will transmit signals from force, pressure and displacement transducers and from closed circuit television cameras set to observe critical points in the model.

Before conducting a test the model should be allowed to come into equilibrium under the increased self-weight stresses at constant centrifuge acceleration; larger models of fine-grained soils may require the centrifuge to be run continuously for several days to reach equilibrium. Often a small ground investigation will be carried out in flight using model cone penetration or shear vane tests similar to those discussed in Sec. 16.5.

A very large number of different events and construction activities can be modelled. Design and manufacture of model loading and construction devices taxes the ingenuity of the engineer and a number of sophisticated and novel examples can be found in the literature of centrifuge modelling. Some typical examples include: vertical and horizontal loading of foundations, piles and anchors; modelling excavation and tunnel construction by draining heavy fluids or by reducing pressures; embankment construction in stages by dropping sand from a hopper; earthquakes simulated by vibrating the base of the model; formation of craters and blast loading on buried structures simulated by detonating small explosive charges.

24.7 SUMMARY

1. Modelling geotechnical structures can be used to examine mechanisms of deformation and collapse, to validate numerical analyses and for parametric studies. Models can occasionally be applied to site-specific cases, but this is usually very difficult.
2. For correct scaling of stresses and soil properties geotechnical models should be tested while under acceleration in a centrifuge. An n-scale model should be tested at an acceleration of ng, where g is the acceleration due to Earth's gravity.
3. At a scale factor of n, rates of loading should be raised by a factor of n and rates of consolidation will be increased n^2 times.

FURTHER READING

Corte, J.-F. (ed.) (1988) *Centrifuge 88*, Balkema, Rotterdam.
Craig, W. H., R. G. James and A. N. Schofield (eds) (1988) *Centrifuges in Soil Mechanics*, Balkema, Rotterdam.
Ko, H.-Y. and F. G. McLean (eds) (1991) *Centrifuge 91*, Balkema, Rotterdam.
Taylor, R. N. (1987) 'Modelling in ground engineering', Chapter 58 in *Geotechnical Engineers Reference Book*, F. G. Bell (ed.), Butterworth, London.
Schofield, A. N. (1980) 'Cambridge geotechnical centrifuge operations', *Geotechnique*, **30**, 227–268.

My objective in writing this book was to set out the basic theories of soil mechanics and geotechnical engineering in a simple and understandable way. In common with introductory texts in other engineering subjects, I have dealt principally with simple idealization to construct a theoretical framework for soil behaviour. You should be aware, however, that this is only a part of the story and the behaviour of natural soils is often more complex.

I have tried to relate the basic principles of soil mechanics to the general theories of mechanics and materials to demonstrate that soil mechanics does have a sound theoretical basis linked to theories that will appear in other courses on structures and fluid mechanics. I have also tried to describe soil behaviour in the context of everyday experiences of the behaviour of soils and granular materials in the garden, on the beach and in the kitchen. I want readers to relate the simple theories of soil mechanics to their own observations. Broadly, the predictions of a theoretical calculation should be what you would reasonably expect to happen and the stability of a large excavation or foundation will be governed by the same theories that govern the behaviour of small holes in the beach.

If you have understood the simple theories in this book, you should be able to analyse a simple retaining wall or foundation and assess the stability of a slope in idealized soil. You should be able to say what soil parameters are required for a particular design, distinguishing between the total stress parameters for undrained loading and effective stress parameters which require knowledge of the pore pressures. You should also know how values of soil parameters for design are determined from ground investigations and laboratory and *in situ* tests and you should have some idea of what are reasonable values for different soils.

Of course, when you graduate you will not be a fully qualified and experienced engineer able to design major groundworks, and the next step in your career may take one of several directions. You might, for example, want to become an accountant, a manager or an inventor and you can do all these in civil engineering. Any construction enterprise is really a business and the engineers will need to manage their resources and account for income and expenditure. Any civil engineering design is really an invention because it is a unique creation and inventors must also be engineers because their inventions must be made to work.

The next step in your career as a civil engineer is to learn how to put theory into practice. You should start by working with experienced engineers and you will be trained through experience. Among other things you will learn how to do routine designs using standard methods. One of the most important things to learn is how to recognize when the problem has become so complex and difficult that you need to consult a specialist.

I hope that some of you will be sufficiently excited by the challenges of soil mechanics and geotechnical engineering to want to become a specialist called on to solve the difficult ground engineering problems. In this case you will probably want to take a higher degree in soil mechanics, geotechnical engineering, engineering geology or a related subject. You will need to know very much more about soil mechanics than I have been able to cover in this book, but it will provide an introduction to these more advanced studies.

The Mechanics of Soils and Foundations will have succeeded in its aims if it conveys to students and engineers the idea that there are relatively simple theories underlying engineering soil behaviour and that form the basis of engineering design. I hope that readers will be able to apply these theories to geotechnical design and use them to assess critically the conventional, routine design methods conventionally used in practice.